T Technik
  Verlag

Busch-Sulzer Bros.-Diesel Engine Company

The Diesel Engine

Technik Verlag

Busch-Sulzer Bros.-Diesel Engine Company

The Diesel Engine

ISBN/EAN: 9783944351018

Auflage: 1

Erscheinungsjahr: 2009

Erscheinungsort: Bremen, Deutschland

@ Technikverlag in Access Verlag GmbH, Fahrenheitstr. 1, 28359 Bremen. Alle Rechte beim Verlag und bei den jeweiligen Lizenzgebern.

GENERAL OFFICES AND WORKS, ST. LOUIS

CONTENTS

Portrait, Dr. Rudolf Diesel	Page 8
Preface	9
First American Built Diesel Engine	10
Historical Sketch	11
Diesel Fuel Consumption, Variable Load	14
Efficiencies of Various Prime Movers	14
Diesel Efficiency and Economy	15
Guarantees	17
Original 225 B.H.P. Diesel Unit	18
Life of the Diesel	19
Diesel Engines in Course of Construction	22
Construction	23
Starting the Diesel	24
Operation	25
Cycle of Operation	26
Combustion	27
Diesel Fuel	28
Advantages of the Diesel	29
Central Station Operation at Light Loads	30
Central Station Practice	31
Flour Mill Drives	34
Plan and Elevation of Company's Engine Room	36
Diesel Power Plant at the Works of the Busch-Sulzer Bros.-Diesel Engine Co., St. Louis	37

Ice and Refrigeration	39
Technical Bulletin on Ice Plants	41
Miscellaneous Diesel Drives	42
Diesel Industrial Applications	43
Exhaust Gas Heat Economizers	46
Representative Diesel Installations	48-88
Service—Visiting Customers' Plants	89
Evidence	90-103
Useful Data and Tables	104-111

D R. RUDOLF DIESEL of Munich, Germany, the distinguished inventor of the Diesel Engine. By agreement with Dr. Diesel this Company has the exclusive right to his services as a director and in a consulting capacity for the United States and Canada. He has given this Company power of attorney to defend the name DIESEL against all infringements.

AN EXTRAORDINARY EFFICIENCY, THE HIGHEST SO FAR KNOWN TO THE ENGINEERING WORLD, AN ABILITY TO ASSUME IMMEDIATELY ANY CHANGE OF LOAD WITHIN ITS CAPACITY AUTOMATICALLY AND WITH PRACTICALLY NO VARIATION IN SPEED, A FUEL CONSUMPTION FROM HALF TO FULL LOAD ALMOST IN DIRECT PROPORTION TO THE LOAD CARRIED, AND AN EXCEEDINGLY SMALL COST OF ATTENDANCE—THESE ARE THE DIESEL'S CLAIMS UPON YOUR SERIOUS CONSIDERATION. SEVENTY THOUSAND HORSE-POWER OF OUR ENGINES IN SUCCESSFUL OPERATION, UNDER BROAD GUARANTEES, IN TWENTY-SIX STATES OF THE UNION, IS A RECORD WHICH SPEAKS WELL FOR OUR AMERICAN DIESEL PRACTICE, AND WHICH ASSURES YOU A QUALITY OF SERVICE SUCH ONLY AS LONG SUCCESSFUL MANUFACTURE CAN GUARANTEE.

THE first American Diesel Engine. Built in Saint Louis and completed September 19, 1898, from designs acquired by Mr. Adolphus Busch, with all American patents and manufacturing rights. It developed sixty horse-power in two cylinders, while the first commercial Diesel built abroad, in the same year, developed but twenty-five in one cylinder. Installed at the Anheuser-Busch Brewery it was the first Diesel to be placed under regular operating conditions.

HISTORICAL

THE Diesel engine was brought to the attention of the engineering world in 1897, when our associate, Dr. Rudolf Diesel, completed his first successful engine at Augsburg, Germany.

Among the first to realize the possibilities of the invention was Mr. Adolphus Busch, who immediately sought the professional advice of Col. E. D. Meier, later president of the American Society of Mechanical Engineers, as to its future.

Mr. Busch and Col. Meier spent several weeks testing the engine at Augsburg, coming to the conclusion that Dr. Diesel's new engine was destined to exert an epoch making influence in the prime mover field, as it showed a thermal efficiency three times that of any steam plant then in operation.

Subsequently a meeting was arranged at Cologne with Dr. Diesel, where a contract was signed which secured to Mr. Busch the entire and complete control of all Dr. Diesel's existing and future patents in the United States, its possessions, and Canada.

Upon Mr. Busch's return to the United States he organized the Diesel Motor Company of America, but this company was soon superseded by the American Diesel Engine Company, it being found that the word motor as used in America was a misnomer when applied to large engines. Both these companies completed a large amount of successful experimental work, endeavoring during this period to maintain correspondence relative to such engineering problems as arose with other lessees then developing the Diesel engine abroad. This correspondence, however, netted practically insignificant results. Our predecessors, therefore, perfected our American type of Diesel to meet the peculiar conditions of American practice, in this manner developing some important elements of design since followed almost universally in Europe.

The first engine constructed under our American rights was completed on September 19th, 1898. It was built in St. Louis and is illustrated on page 10. It operated in the Anheuser-Busch Brewery until superseded by larger units.

In February, 1911, Mr. Busch, who had become the purchaser of the American Diesel Engine Company organized the present Company, the Busch-Sulzer Bros.-Diesel Engine Co., thus associating his Diesel interests with the Gebrüder Sulzer, of Winterthur, Switzerland. The Gebrüder Sulzer are recognized as among the foremost builders of high-class machinery in the world, and without a peer in the development of the Diesel engine.

By this sagacious move Mr. Busch realized his ambition to combine the long experience and only experience of Diesel building in America with the best that Europe affords in Diesel engineering, manufacturing and experience; Sulzer Brothers and Dr. Diesel thus becoming interested, both financially and as directors, with Mr. Busch.

Our St. Louis plant, representing an investment of a million dollars, is equipped with every device and convenience for the proper handling of Diesel manufacture, according to the most approved modern practice.

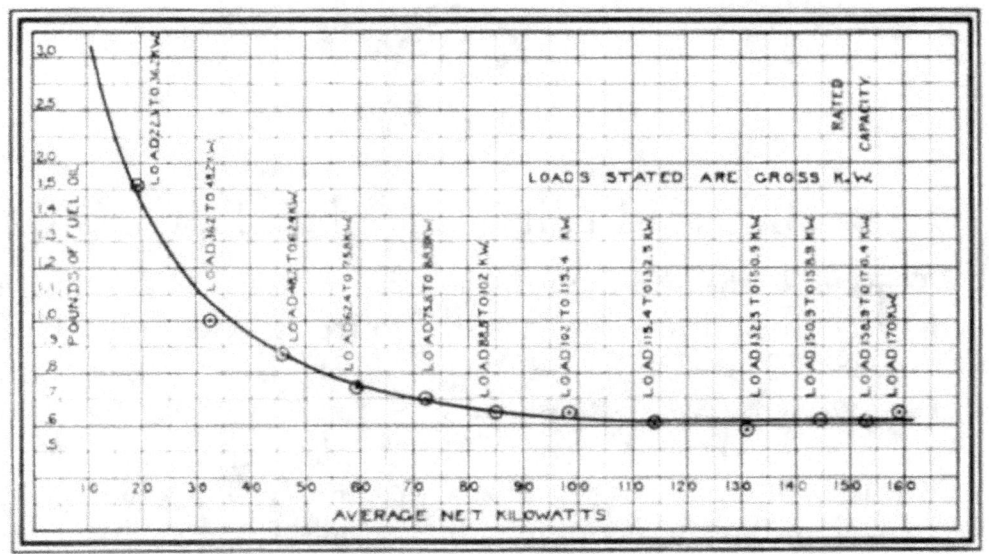

DIESEL FUEL CONSUMPTION PER K. W. HOUR AT VARIOUS LOADS

THIS curve shows the results obtained under actual working conditions, and brings out the remarkable maintenance of efficiency, from full load down to half load, which is the unique characteristic of the Diesel Engine.

❖ ❖ ❖

PRIME MOVER EFFICIENCIES

Type of Plant	B.T.U. per B.H.P. Hour	Efficiency
Non-Condensing Steam Engine	30,000-38,000	8.4%-6.6%
Condensing Steam Engines and Turbines using superheated steam at 150 pounds	17,000-25,000	15%-10%
Diesel Engines	7,500- 8,000	35%-32%

One H. P. Hour $\frac{33,000 \times 60}{778 \text{ B.T.U.}}$ 2545 B.T.U. per Hour

DIESEL EFFICIENCY AND ECONOMY

THE thermo-dynamic efficiency of the Diesel Engine, based on net useful output, varies between 32 per cent. and 35 per cent.; that of the simple Corliss or 4-valve engine is about 6 per cent.; the Corliss compounded, 9 per cent.; the triple expansion engine, rarely 18 per cent. The very high efficiency of the Diesel makes it economically possible to purchase more expensive fuel than for steaming, and still show a handsome profit by its operation—and this without the necessity of considering its other advantages, which in many cases are as important as its extraordinary fuel economy.

Diesel economy of space, fuel and attendance; its elimination of all stand-by expense; its fuel consumption from half load to 10 per cent. overload, almost in direct proportion to the load carried; and its readiness to start cold at a moment's notice—these are responsible for its unprecedented efficiency and magnificent economy.

Diesel engines eliminate coal bunkers, stacks, boiler room and boiler room auxiliaries. They eliminate incompetent and careless stoking, firing, draft and water regulation—losses which, even in well regulated steam plants commonly amount to from 15 to 30 per cent. the value of the coal. They eliminate the varying factors to which coal itself is subject—its varying percentages of moisture, ash and oxygen; also calorific deterioration due to storage, which in half a year may amount to 12 per cent.—changes in composition which require careful changes in handling, if efficient combustion is to be approximated.

The Diesel uses less water than required for the operation of producer gas engines or condensing steam plants of like power.

A shortage of motive fluid, which sometimes occurs in steam plants, due to unexpected increases in load, where requirements cannot be anticipated, is a failing unknown to Diesel installations. No such shortage is possible where Diesels are installed. Furthermore, it is not possible to waste Diesel fuel through an unlooked for return to lighter loads.

We guarantee the economy of the Diesel under everyday commercial conditions, although no builder of steam engines or accessory equipment will guarantee *his* product either as to steam or fuel consumption, except for brief full-load tests under exact specific limitations. It is an inherent failing of steam plants that they have factors which must be left to the discretion of attendants, factors which can be judged only by men skilled in analyzing temperature and draft records, flue gas, coal and ash analyses, water consumption, etc., etc.—variable factors to be constantly and intelligently analyzed if steaming efficiency is to be approximated. Diesel economy is not dependent on ceaseless vigilance and unerring judgment. It is controlled solely by means of a sensitive governor by which the rate of fuel injection is instantly modified to meet momentary load requirements. This it accomplishes with such precision that the conditions of parallel operation of alternators is controlled solely and perfectly by the regulation of fuel consumption.

We take pride in giving here an example of the remarkable fuel saving of our Diesel engines:

We have installed a plant in the middle west, consisting of two units of 240 B.H.P. each, which is saving its owner $10,500 a year in fuel or more than $30 per brake horse-power-year over the fuel cost of the old superseded steam equipment, which consisted of a condensing Corliss engine. This Diesel installation carries a steady, heavy mill drag twenty-four hours daily, together with the frequent intermittent service of an elevator of 350,000 bushels capacity. We refer the reader to section entitled "EVIDENCE" for numerous other examples.

GUARANTEES

THE Company guarantees the Diesel against defective parts due to faulty material or workmanship, and will replace such parts free of charge.

The Company guarantees that the variation in speed of its engines will come within the close limits required for the parallel operation of 60-cycle alternating current generators.

The Company further guarantees the engines as to economy of fuel consumption.

Copies of the guarantees which this Company offers and the tests by which the same are demonstrated will be cheerfully furnished to prospective purchasers.

Every engine, before shipment, is carefully and thoroughly tested in our shops, at one-quarter, one-half, three-quarters, full and overload; and a certified copy of the results is furnished with each engine.

See sections of this general catalogue entitled "REPRESENTATIVE DIESEL INSTALLATIONS" and "EVIDENCE" for records which Diesels are making under everyday operating conditions.

ONE of our original 225 B.H.P. Diesel Units. In economical operation since 1904.

LIFE OF THE DIESEL

IT is not unusual to find steam engines and steam pumps which have been in service thirty years, maintained in fair state of up-keep by repairs and renewals—the frame, shaft, flywheel and foundation, representing a large part of their original cost, continuing in service. However, the engine and pumps represent less than forty per cent. of the cost of a steam installation—approximately sixty per cent. being in the boilers, heaters, condensers, stack and piping. Some of these features, the boilers notably, each year show a marked deterioration and loss in efficiency. None of these features exist in the Diesel, and its life will compare most favorably with the entire equipment of a steam plant, its efficiency throughout its life remaining practically unimpaired.

The story of the Diesel Engine is quite different from that of gradual obsolescence of the old steam plant. Ten, even fifteen years ago, when the Diesel was first built, it showed the same extraordinary efficiency. No builder of

Diesels abroad, nor do we here, expect to increase its thermal efficiency to a very great extent. Diesel progress has been one of increasing refinements, a lengthening of its life, an increasing of its reliability and facility in handling, in its close governing under varying loads, etc. In these it is unapproached by any other type of prime mover.

The heavily designed frame, the shaft, and connecting rods, the massive fly wheel, etc., form a much larger proportionate cost of Diesel equipment than these parts do in a steam installation, and since these non-wearing parts form the larger cost, those parts which wear and deteriorate most, of necessity, form the smaller and a lesser proportionate part of Diesel equipment than they do with steam equipment. It is easy to realize this—if one will recall that the entire boiler equipment with all its auxiliaries is eliminated, and that wear and tear is confined to parts which represent less than one-third of the original Diesel investment.

In the steam engine and in all explosive and hot-bulb types of internal combustion engines, leaky valves and worn cylinders result in reduced efficiency, the cause of which is not always apparent, and if the engine is not loaded to capacity may not be detected until much damage has been done and much money lost in poor efficiency. The Diesel, depending upon perfect compression for its ignition, does not permit a continuance of such losses; if compression fails ignition ceases and the engine stops. In other words such conditions as militate against the life of engines and their economy absolutely cannot exist long enough in the Diesel to do serious damage, or eat up fuel in useless effort.

Another feature of the Diesel which adds to its life, and which sets the Diesel apart from all explosive types, is the absence of any sudden rise in pressure at instant of combustion. Gradual introduction of fuel during ten per cent. to twelve per cent. of the combustion stroke results in a more uniform stress and longer life.

There are two 225 B.H.P. Diesel engines in a Texas power house, installed nine years, during which period they have operated on an average eighteen hours per day. Cylinders of these engines have never been rebored, show negligible wear and are smooth and bright as glass. With the same handling in the future as they have had in the past, they should outlive a steam plant of like capacity.

DIESEL ENGINES IN COURSE OF CONSTRUCTION

CONSTRUCTION

FIFTEEN years of Diesel building have shown us one conspicuous fact in relation to construction which we deem fundamental; that if a Diesel were built and operated under average conditions, with no more care given to material and construction than is usual in steam engine practice, Diesel efficiency would be greatly impaired and operation would not be reliable. Our close scrutiny of details, and our strict adherence to the highest type of engineering are responsible for the success which has attended the type constructed by this Company.

We commenced building Diesel engines in 1898, in the same year commercial development began in Europe, and have since given our best attention to the perfection of a design in consonance with American practice which would embody all those features found by experience to increase the remarkable reliability of Diesel operation.

Our type, characterized by compact simplicity of design, embodies great convenience with highest efficiency. Our methods assure perfect interchangeability of parts, all of which are liberally proportioned, with workmanship, material and design standardized and in strict conformity with our general practice, determined by an experience in Diesel building extending from its introduction to the present time.

STARTING the Diesel in an United States Naval Torpedo Station. A twist of the wrist does it. In less than three minutes a Diesel will take on full load.

OPERATION

THE Diesel Engine, if designed and built in accordance with the lessons of practical experience is absolutely dependable for the severest service and the longest non-stop operation. Our customers operate Diesel Engines over regular periods of six weeks to two months without shut-down. They operate them without realignment or other major adjustment for periods of years. Even for the severest service our Diesel engines require less attendance than any other type of prime mover.

The duties of attendance during operating periods consist principally of watching lubrication, seeing that the flow of cooling water is uninterrupted and in keeping the engine clean. A first-class mechanic or steam engineer is amply qualified for this service and may be easily trained to operate the Diesel intelligently. The various duties during shut-down periods should be divided between examination and adjustment. Periodic inspections should occur at regular intervals more or less frequent, depending upon the severity of the service. The actual work involves grinding valves, adjusting boxes, packing glands, and the renewal of lubricating oil—the same sort of duties found in every steam plant. There are, of course, no boiler tubes to replace, boiler scale to remove, flues to clean, heat insulation and grates to renew, brick work to be patched or the like. So that, more than anything else, Diesel operation and attendance mean watchfulness, as there is an almost complete elimination of manual effort. As it is with steam, procrastination is the root of most trouble, and the test of the fitness of an operating engineer.

The Stationary Diesel which we build belongs to the four-stroke cycle type of internal combustion engine, the cycle of each cylinder being completed in two revolutions of the crank or four strokes of the piston;

(first) INDUCTION of pure air, (second) COMPRESSION of pure air, (third) COMBUSTION of oil sprayed in the compressed air, and EXPANSION of the products of this combustion; (fourth) EXPULSION of exhaust gasses.

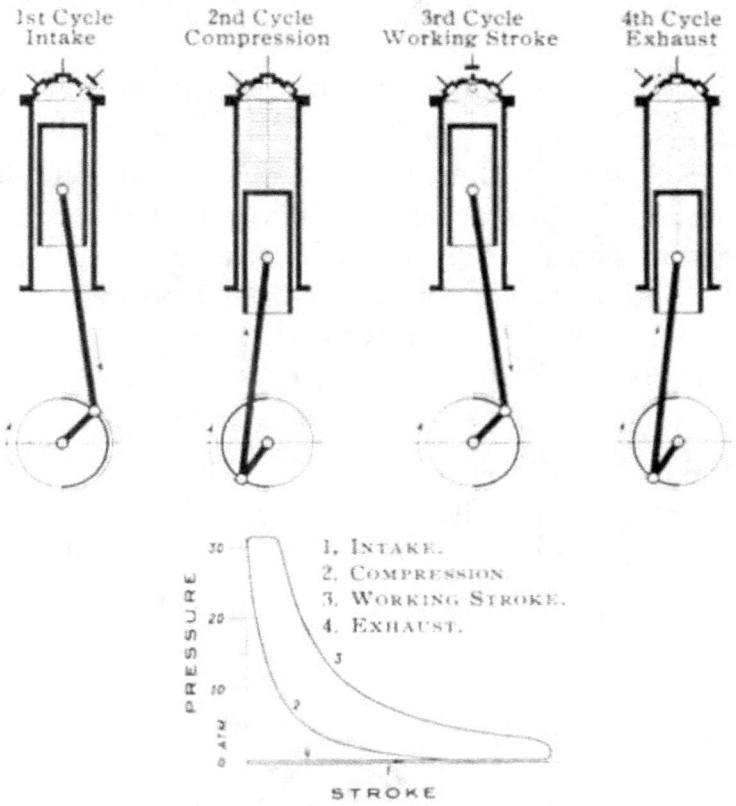

The Diesel does not contain an explosive mixture at any time, no explosion ever occurs in its cycle of operation, and the Diesel never was and never will be subject to pre-ignition, as air only is compressed. No carburetor, no vaporizer, no hot-bulb, flame, or electrical ignition apparatus is ever used. Combustion of the oil spray is due solely to the heat generated by compression on the second (COMPRESSION) stroke of the cycle. The spraying of the oil into the cylinder covers from 10 per cent. to 12 per cent. of the COMBUSTION stroke. It is a gradual burning, continuing for a considerable time after all the fuel has been injected—a non-explosive, internal combustion resulting in uniform stress and long life—the Diesel is the only engine which has it.

COMBUSTION

CRUDE Oil, or residuum, commonly known as fuel oil, burning temperature 120° to 300° Fahrenheit, forced through an atomizer by an air blast, enters the combustion space of the cylinder at the point of highest compression when the air, drawn in on the suction stroke, has been compressed to 460 pounds and thereby raised in temperature to 1000° F. This is a temperature 3 to 8 times that required for ignition. Instant combustion follows, and every combustible particle burnt.

Diesel combustion is combustion in incandescent atmosphere under ideal conditions insuring perfect combustion, smokeless exhaust, and the highest thermal efficiency known.

DIESEL FUEL

CHEAP fuel oils containing non-combustible substances, or high percentages of sulphur are not always the most economical Diesel fuels. Such oils are bad for all internal combustion engines regardless of their type or design, although there are sometimes market conditions under which they may be used profitably.

Will the saving, amounting to the difference in cost between such oils and those free from such impurities, warrant the cost of frequent replacements of those parts attacked by the sulphur and worn by the non-combustible matter which the cheaper contain? This is the sum and substance of the fuel problem, and varies in no respect, except in degree, from that confronting every plant manager, no matter what type of prime mover he may operate. As to degree—coals for steaming vary greatly in heat units per pound, cost of handling, etc.; oils for hot bulb and explosion type engines are available only between certain narrow limits; as to oils for the Diesel, there is the greatest latitude in choice, oils from practically all fields having been used successfully, their thermal value never entering as a factor in their purchase or cost. Unlike coal, which has a calorific value ranging from 8,000 to 14,500, all heavy oils, such as Diesels consume, have approximately the same high heat value—namely 19,000 B.T.U. per pound.

The Company will advise its customers and interested inquirers as to the availability of any particular oil and invites their correspondence on this subject.

ADVANTAGES

- No Boilers.
- No Boiler Explosions.
- " " Inspection.
- " " Cleaning.
- " " Repairs.
- " Engineers' licenses.
- " Firemen.
- " Smoke Nuisance.
- " Preliminary Heating-up.
- " Banked Fires.
- " Ashes.
- " Dirt.
- " Dust.
- " Shut-down for coal strikes.
- " Stand-by losses.
- " Over-heated buildings.
- Fuel easily handled and stored.
- Larger quantities of reserve fuel easily stored.
- No depreciation on stored fuel.
- No losses of fuel in transit.
- Less water wasted.
- No ignition troubles.
- Never fails to start.
- No explosions or sudden shocks.
- Absence of lubrication difficulties.
- Less floor space required.
- Power available immediately.
- Perfect regulation.
- Great economy of operation.
- Highest thermal efficiency known.
- Practically same economy at half load as at full load.
- Long non-stop operation.
- Absolute dependability.

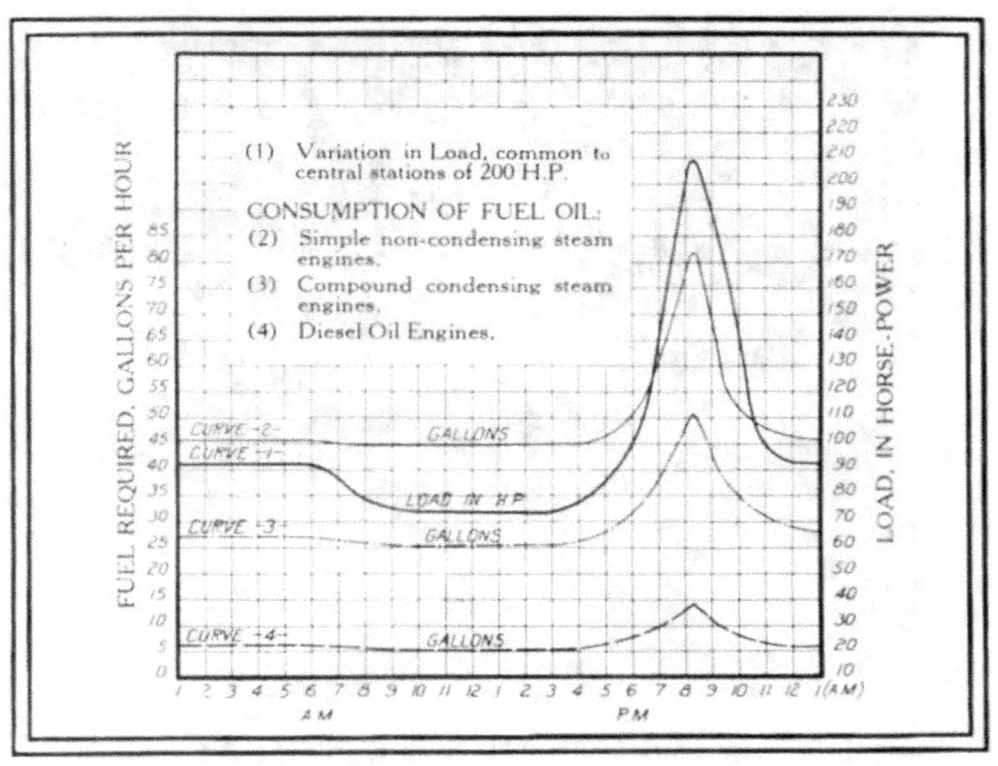

CENTRAL STATION OPERATION AT LIGHT LOADS

THIS chart shows the range of load common to Central Stations. It shows that where the Diesel consumes five barrels of oil the condensing steam engine consumes twenty-five barrels and the simple steam engine forty-five barrels. This gives an economy in favor of the Diesel of one to five and one to nine.

The chart shows a day load of 75 H.P. rising at dusk to a peak of approximately 200 H.P. falling to an after midnight load of 93 H.P. The Diesel carries this day load of 75 H.P. on six gallons of oil per hour. Ask your engineer to compare this record with your results. How much fuel do you use per hour during off-peak periods? How much per day?

CENTRAL STATION PRACTICE

IN December, 1907, Diesel engines to the amount of 9,665 brake horse-power were operating in Central Station plants in the United States. Of the plants operating this Diesel power, 75 per cent. have bought additional units, their re-orders, in brake horse-power, amounting to 155 per cent. of the original amount purchased on first order, 55 per cent. re-ordering after original purchases had shown five years or more of successful, economical operation.

The horse-power sold on re-order to these central station plants now amounts to 68 per cent. of all the horse-power they had in operation in 1907. That sold on re-order up to and including December, 1907, was 29 per cent. of that installed at the time; while that bought on re-order at this date is 38 per cent. of that now in operation. Smallest equipment operated in a Diesel central station is of 75 B.H.P., largest 1125 B.H.P.

All the above figures refer to strictly central stations deriving all income from such service. This Company has equipped central stations of more than double the size of largest indicated above, but which are engaged in other lines also, as in manufacturing, mining and street railway operation—all of which have ordered additional Diesel units.

Of those which have re-ordered, 24 per cent. have made three distinct purchases, each in a different year, and one power and light company, with Diesels installed in two of their plants, has made four purchases:

        450 B.H.P. in 1904
        450 B.H.P. in 1910
        225 B.H.P. in 1911
        225 B.H.P. in 1912

These repeat orders over so many years, show that the Diesel is well adapted to Central Station requirements of regulation, reliability and continuous operation.

PARALLEL OPERATION

Diesel speed regulation under change of load ranks with that of the best types of automatic steam engines. No difficulty is experienced in the operation of generating units in parallel. Reference can be given to large numbers of such plants operating in parallel with other Diesel units, steam equipment and water power, in different parts of the country.

RELIABILITY AND CONTINUOUS OPERATION

In the Diesel Engine, combustion by the heat of compression does away with ignition devices, mixers, carburetors and back firing, limiting the cause of stoppage to a cessation of fuel or injection air. Any part working out of adjustment gives such ample notice of a fault, that, generally, attenion may be deferred to regular shut-down periods.

One 225 horse-power Diesel Engine installed in an electric light plant in Illinois, operated without reserve power, 24 hours per day—$6\frac{3}{4}$ days per week—for $2\frac{1}{2}$ years with but two minor shut-downs. In the opinion of operating engineers, who have had several years experience, the Diesel is fully as reliable as steam.

ECONOMY IN SMALL SIZES AND UNDER VARIABLE LOAD CONDITIONS

The refinements of coal handling machinery, superheaters, economizers, and labor saving devices commonly found in large modern steam plants are not economically introduced into small central stations, where the varying load condition is most marked. Therefore, the kilowatt costs several times as much in the small steam installation as in the large well equipped station. In contrast to this the small Diesel installation shows a kilowatt cost which compares very favorably with that obtained in the larger plants operated by Diesels or the most refined steam equipment.

The installation of small Diesel units allows a gradual increase in capacity to meet growing load conditions, and has the additional advantage that the factor of safety for continuous operation increases with the number of units installed—any necessity for repairs or adjustments affecting a smaller percentage of the total capacity of the plant.

FLOUR MILL DRIVES

DIESELS are installed in flour mills which have both electrical and line shaft drives. With both types Diesel Engines are showing an economy which has amply justified their installation.

A mill and elevator company operating a 320 H.P. and an 80 H.P. Corliss, both running condensing, changed over to Diesel equipment. An 800 barrel mill is now driven by a 250 H.P. motor, a 400 barrel mill by a 125 H.P. motor, the cleaning machines by a 50 H.P. motor, and the elevator by 7 motors aggregating 200 H.P.

This mill, located in the middle west, burned an average of 55 barrels of oil per day under boilers, the lowest consumption recorded having been 48 barrels. It is now consuming in its Diesel engines an average of 12 barrels. It cost this mill more than $4\frac{1}{2}$ times as much to run formerly as now—for fuel alone. Add to this great Diesel economy in fuel, the saving in labor, and the showing is well worth investigating. The owner of this mill will tell you that he is saving over $10,500.00 per year in fuel alone.

Figuring the oil at $2\frac{1}{2}$ cents per gallon, the local price, the fuel cost is only $10\frac{1}{2}$ mills per barrel of flour. As the efficiency of the smallest Diesel more nearly approaches that of the largest, than does the efficiency of a small steam plant that of a large one, a flour mill requiring only 120 B.H.P., Diesel operated, would show a proportionately greater saving.

At the head of this article is a photograph of the Rope Drive of a recent Diesel Flour Mill installation in Texas.

The first Diesel to be used in a flour mill was installed in 1905, and has been in successful economical operation ever since. This engine is clutch-connected to line shaft. In the REPRESENTATIVE DIESEL INSTALLATIONS section, under Kansas and Texas, two roller mill installations will be found illustrated.

If you have a mill drive write us for further information on the availability of the Diesel in your mill.

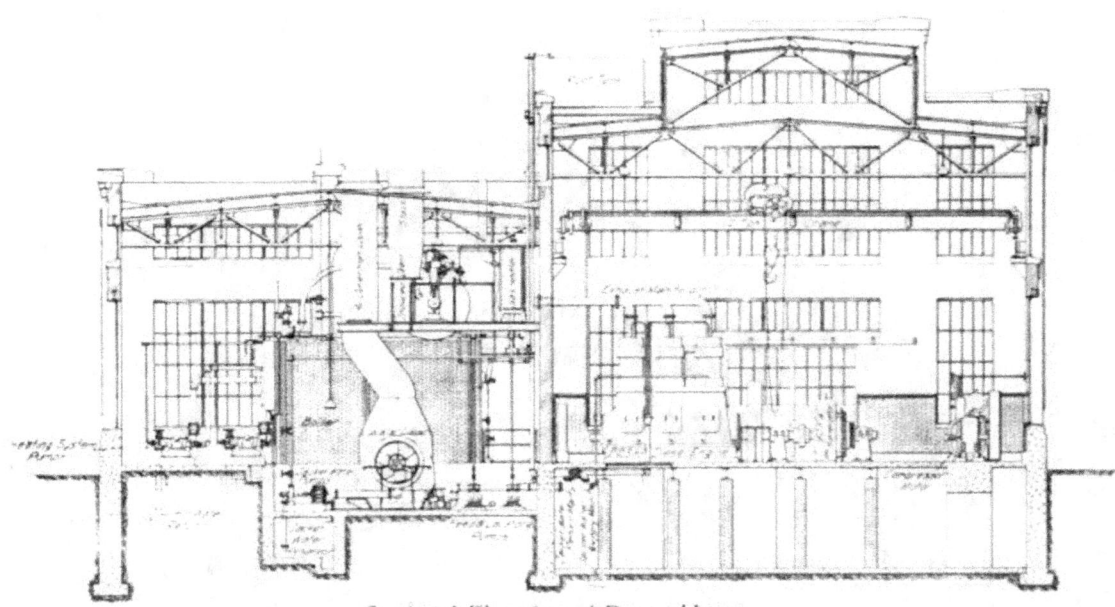
Sectional Elevation of Power House.

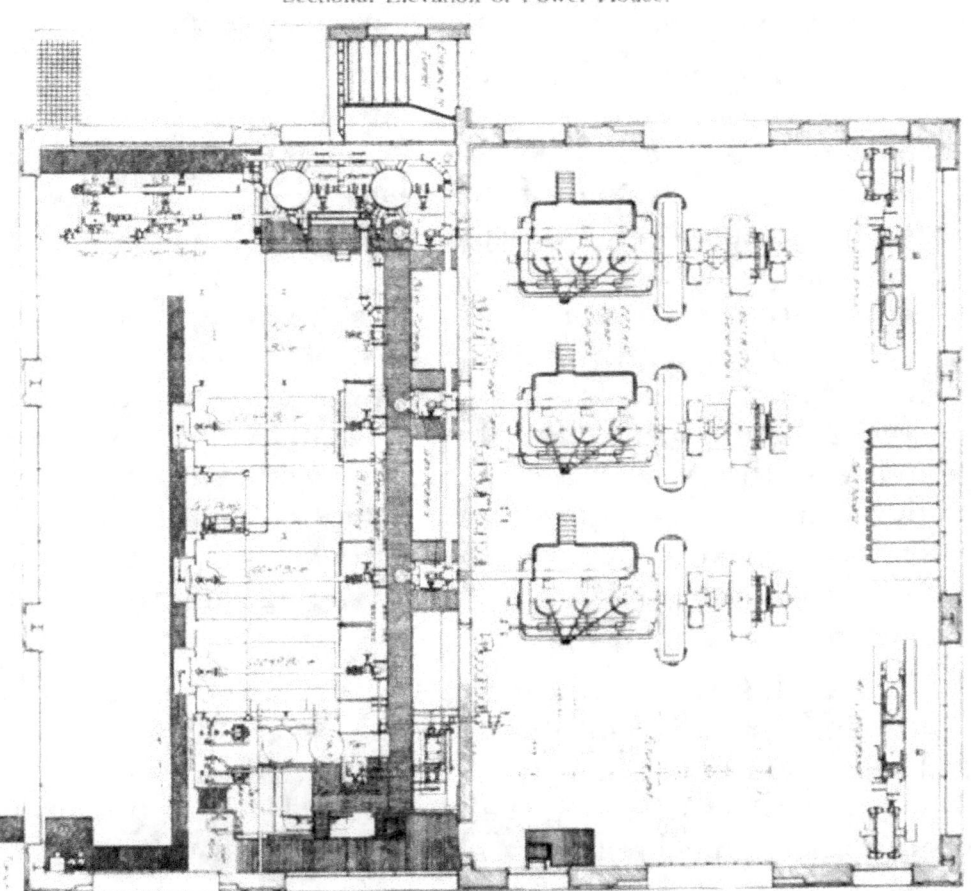
Plan of Power House.

PLAN AND ELEVATION OF THE COMPANY'S POWER HOUSE.

DIESEL POWER PLANT
AT THE WORKS OF THE
BUSCH-SULZER BROS.-DIESEL ENGINE CO.

ALTHOUGH the Diesel turns to useful account approximately twice as much of the heat value of fuel as do steam engines or turbines, yet, with about 32 per cent. to 35 per cent. of the heat transformed into useful mechanical energy, there remains 43 to 40 per cent. not available to that purpose—turned into heat.

In this installation the exhaust gases from the Diesels pass through exhaust gas heat economizers (illustrated and described on page 46). The returns from the hot water heating system of this plant pass through three of them, arranged in parallel, one for each engine. After absorbing the heat of the gases, with a consequent rise in temperature, the water passes from them to heaters which are supplied with steam coils heated by the steam of high pressure boilers, used also for the operation of the steam hammers in the forge shop. From the steam heaters the water again passes out and re-circulates through the buildings. The cooling water from the cylinder jackets, at a temperature of about 140 degrees, runs into a hot well, this water being utilized as boiler feed and in lavatories.

In this manner the system employed in this plant, the first in this country to utilize heat from both these Diesel sources, conserves to heating purposes at least 60 per cent. of the heat value of the fuel not transformed into mechanical effort. The result is a very decided economy in fuel for winter heating, an economy which, when added to that of the Diesel as a prime mover, makes for an overall fuel economy which is superb.

Another important economic feature of this plant is automatic machine tool control. While this is not a feature which can be used only with Diesel engines, individual motor drive with automatic control is, however, one of those important items subsidiary to the Diesel making for general manufacturing economy. It is a system which necessarily has a maximum variation in load, as current is consumed only during periods of actual productive work. Constant voltage is its only requisite to successful operation, and this is easily provided by the Diesel, any increase or decrease in load being immediately reflected in a changed rate of fuel consumption, always in proportion to the productive work being done.

The Diesel engine equipment of this plant consists of three 4-stroke cycle Diesel units of 225 B.H.P. each, running at 164 R.P.M., direct connected to 160 K.W. 115-230 volt direct current generators. The fuel oil storage tanks, of which there are two, are buried at the side of a private railroad siding adjacent to the power house. Oil flows into these tanks from tank cars by gravity. Oil fired boilers are used in the power house—they supply the major portion of the heat required for shop heating in the winter and the steam for the operation of the steam hammers in the forge shop.

Diesel exhaust is noiseless, colorless and odorless, and as induced draft is used in conjunction with the boilers this installation is without visible stacks of any kind. No coal is used on the premises. Even in the forge shop oil has been substituted throughout on account of its ease in handling, its economy and the precision of its control.

This company considers that the use of the Diesel engine, together with such heat conserving equipment as is employed in this plant, assures to all plants having a heating problem which will install Diesels, an overall economy which cannot be approached by any other type of prime mover.

ICE AND REFRIGERATION

WE have installed Diesels in a number of ice and refrigeration plants. Some are operated in conjunction with electric light stations, some with water works, some with breweries, some as distinct plants. These Diesels are connected to load by belted and electrical drives and are particularly well adapted to ice manufacture, showing a reduction in fuel cost of 40 per cent. to 80 per cent. and an operating expense closely proportioned to output.

Inasmuch as the fuel consumption of the Diesel is in direct proportion to load requirements, between half and full load, it follows that, with one Diesel engine, the same fuel cost per ton of ice will be realized at half as at full capacity; with two Diesels, at one quarter to full capacity, etc. This flexibility is appreciated in off seasons. In the summer when ice making continues twenty-four hours daily the Diesel is reliable for full capacity, twenty-four hours per day, for the full season, with one or two shut downs for inspection and possible adjustments.

Six years ago when an ice plant in the south installed its first Diesel engine it was consuming under its boilers an average of $500 worth of fuel per month. One year later, after its second Diesel was installed and the steam plant abandoned, and the business had increased, the fuel consumption averaged only $75 monthly. The third Diesel this company purchased was installed in 1912, five years after they had bought their first, ample time for them to have discovered whether or not the Diesel fully met the requirements of ice and refrigeration service. Their Diesels operate two 125 K.W. and one 150 K.W., A. C. generators of 2300 volts, which are direct connected. Their equipment, including the forty ton ice plant, is operated on twenty-four hour service by one chief engineer and two assistant engineers—more help not required. The Diesel enabled this company to produce a raw water can ice which their competitors could not equal and which captured every retail dealer in town, forcing their rivals to retail their own ice. This plant produces cakes weighing over six hundred pounds without flaw or blemish which are described as blocks of crystal.

Two views from the largest Diesel ice and refrigeration plant, which is situated in New York, are reproduced herewith. It is equipped with six Diesels aggregating 1245 B.H.P. and has a capacity equivalent to 455 tons of ice; its ice machines, pumps and hoists being operated by motors.

SEND FOR TECHNICAL BULLETIN ON DIESEL OPERATED ICE PLANTS

A PAPER was read before the International Congress of Ice and Refrigeration held at Chicago, September 17-25, 1913, written by Messrs. R. H. Tait and L. C. Nordmeyer of the firm of Tait-Nordmeyer Engineering Company, Saint Louis. It has been recast for us by the authors—use being made of the parallel column method of comparison in such manner as to insure the most vivid and ready conception of their estimates, the calculations by which they figured them and the conclusions they reached. It is a technical bulletin on the use of the Diesel Engine in ice plants. We will cheerfully send this bulletin to anyone with a power problem in his ice or refrigerating plant. It contrasts simple steam, compound condensing steam, and Diesel engined 60 ton ice and 12 ton refrigeration capacity plants. It gives the building and operating costs of each and other valuable information to the man who is confronted with a power problem.

MISCELLANEOUS DIESEL DRIVES

ENTIRELY aside from the great economic advantage of the Diesel, it is wonderfully well adapted to high pressure fire service, or any other emergency service which requires instant readiness to start, ability to make long non-stop runs, and absolute reliability. In such plants stand-by expense is often the largest single item and the Diesel eliminates it. If prepared for starting, when shut down, the Diesel can take on full load in less than one minute from notice to start. These characteristics entitle it to the most serious consideration of all having such requirements.

In the foregoing pages there has been described the economic advantages of the Diesel in central stations, refrigeration plants, flour mills and factories. The horsepower thus employed represents 76 per cent. of the total installed.

On the following pages we mention a number of different lines of industry in which Diesels have been running over a period of years—they will give you an idea as to the great range of application to which they have been put.

Our Sales and Engineering Departments would take pleasure in giving a prospective power purchaser all the special information within their experience as to the Diesel's availability under any specific conditions.

DIESEL INDUSTRIAL APPLICATIONS

FOLLOWING is a list of installations illustrating the Diesel's applicability to various lines of industry. A representative installation of either the oldest or largest of each industry is here referred to by its date of installation and location by state. Thirty per cent. of the total Diesel horse-power sold has been on re-order; and the re-orders have amounted to more than the original purchases. On an average those firms which have added Diesels to their first purchases, have more than doubled their plants by fifteen per cent. One central station plant has re-ordered three times, six years after the original purchase, one year after the second, and one year after the third, making four distinct purchases in eight years, which clearly indicates the satisfaction which the Diesel gives to those who have come to know it well. All installations here referred to have been in successful operation since the date mentioned.

Automobile Works	1907	Indiana
Auxiliary to Water Power	1910	Minnesota
Battle Ships	1912	Argentine
Battery Charging	1903	New York
Brewery	1898	Missouri
Cement Works	1912	Oklahoma
Chocolate Works	1911	Pennsylvania
Copper Mining	1906	Arizona
Cordage Works	1906	Massachusetts
Correctional Institution	1912	Pennsylvania
Cotton Gin	1913	Texas
Cotton Goods	1907	Massachusetts
Cotton Seed Oil	1913	Texas
Cotton Twine	1906	Massachusetts
Cotton Webbing	1906	Massachusetts
Drop Forge Shop	1906	Rhode Island
Electric Light Plant	1903	Florida
Encaustic Tiling	1912	Ohio
Fertilizer Works	1912	Massachusetts
Flour Mill and Elevator	1905	Texas
Foundry and Machine	1905	New York
Gas Works	1907	Pennsylvania
Glass Works	1907	Indiana
Gold Mines	1906	New Mexico
Ice and Cold Storage	1904	Texas
Ice Cream	1913	Texas
Iron Doors, Windows	1903	New Jersey
Isolated Plant	1903	Illinois
Knitting Mill	1910	New York
Locomotive Works	1905	Pennsylvania
Machine Builders	1902	Rhode Island
Machine Shop	1902	Rhode Island
Marble Works	1904	New York
Municipal Plant	1903	Florida
Naval Torpedo Station	1911	Rhode Island
Navy Yards	1913	Mass. and Hawaii
Nut and Bolt	1907	Rhode Island
Office Building	1905	Indiana
Optical Works	1907	Rhode Island
Ordnance	1902	Connecticut
Paper Mills	1906	Indiana

Phosphate Mining	1906	Florida
Piano Factory	1905	Maryland
Printing Ink	1912	New Jersey
Quarries	1905	Connecticut
Reduction, Metallurgical	1907	Arizona
Roller Mills	1913	Kansas
Railroad Shops	1907	Texas
Ry. Tunnel Ventilation	1913	Illinois
Sheeting, Cotton	1910	Massachusetts
Silverware	1905	Rhode Island
Street Railway	1904	Texas
Structural Steel and Iron	1904	Indiana
Submarine—Sulzer Engines		
Time Recording Instruments	1905	New York
Waterworks	1905	Wisconsin
Wholesale Grocery	1903	Illinois
Wire Mills	1907	Rhode Island
Woolen Mills	1905	Massachusetts
Worsted Mills	1905	New Jersey
Yarn	1909	Massachusetts

Eighteen states are included in this representative list of sixty-one plants in as many different lines of industry and application, but there are fine installations in eight more states—Arkansas, Iowa, Louisiana, Missouri, New Hampshire, South Carolina, South Dakota and Tennessee —some of which are illustrated elsewhere in this catalogue.

RIVADAVIA-Argentine. Largest and swiftest battleship afloat, equipped with Sulzer-Diesel Marine Auxiliaries furnished by this Company.

EXHAUST GAS HEAT ECONOMIZERS

EXHAUST Gas Heat Economizers which extract the heat of exhaust gases and transfer the same to water find economical uses in such industries as require hot water, as in the washing of raw materials or finished products, in hot water heating systems for buildings and shops, in lavatories, or as boiler feed. At small expense for the equipment and its installation these economizers in conjunction with the engine jackets, will save at least sixty per cent. of the heat which would otherwise be wasted, or 2800 British thermal units per brake horse-power-hour, giving the Diesel an unapproached overall economy.

The following figures are taken from a test in a Sulzer built installation of 300 B.H.P. in a woolen mill at Bürglen, Switzerland, where the heat recovered is used in heating the factory and in the washing of wool and yarns. The tests were made by Prof. J. Cochand, of Lausanne and Engineer M. Hottinger, of Winterthur, Switzerland, their report being published in the Zeitschrift des Vereines Deutscher Ingenieure, of March 23rd, 1912.

Two economizers of 325 square feet heating surface each were used, connected in series so that the exhaust traveled through one then the other. The water, on leaving the engine, passed through them on the counter current principle, in the opposite direction to the flow of the gases. 1232 gallons of water passed through the heaters per hour entering at 123.8 degrees F. and leaving at 167.9 degrees F. The temperature of the water as it entered the cylinder jackets—70 degrees F.

300 B.H.P. PER HOUR

B.T.U. transformed into horse-power	731,274	33.5%
B.T.U. recovered in cooling water	596,691	27.4%
B.T.U. recovered by economizers	460,917	21.1%
B.T.U. lost	393,824	18.0%
B.T.U. in fuel used per hour	2,182,706	100.0%

Deducting the 18 per cent. lost this gives an overall fuel economy of 82 per cent., which is not equalled by any other type of prime mover. As the amount of heat recovered by the economizers and from the cooling water amounts to 48.5 per cent. it may be readily seen that for every $1,000 spent in Diesel fuel but $515.00 worth can be charged to the generation of power. $485.00 worth, or that saved from these sources is utilized in the form of heated water. 48.5 per cent. is a large saving, especially so, considering the small cost of the equipment. But the saving is really much greater. $485.00 worth of oil burnt under a boiler would not come within 20 per cent. to 25 per cent. of the heat recovered and utilized by the exhaust gas heat economizers of this Bürglen plant.

This Company employs this system of heat conservation in its Saint Louis works and is prepared to contract for similar installations in conjunction with all Diesel engined plants.

REPRESENTATIVE DIESEL INSTALLATIONS

FORTY INSTALLATIONS IN TWENTY-THREE STATES ARE REPRESENTED IN THE FOLLOWING VIEWS OF DIESEL POWER PLANTS. SPECIFIC INFORMATION CONCERNING OUR NUMEROUS INSTALLATIONS FURNISHED PROSPECTIVE PURCHASERS ON APPLICATION.

IN ONE PART OF THE COUNTRY OUR DIESELS ARE LOCATED IN MINES; IN ANOTHER, FACTORIES AND CENTRAL STATIONS; IN ANOTHER, AS AUXILIARIES TO WATER POWER—ONE SELLING ANOTHER. DIESEL APPLICATIONS COVER FIFTY-EIGHT DIFFERENT LINES OF INDUSTRY.

IN EVERY LOCALITY WHERE SOLD, INSTALLATIONS HAVE MULTIPLIED. DIESEL POWER PLANTS, SUCCESSFUL AND ECONOMICAL IN OPERATION, ARE OUR BEST SALES AGENTS.

ARIZONA

THIS double unit consists of two 225 B.H.P. Diesel engines. It is located at a copper mine at an elevation of 5,000 feet. It is connected by rope drive to a positive pressure blower. It is run continuously twenty-four hours daily and is not shut down for months at a time, and then only when adjustment is imperative. As the cessation of air to a smelter would result in great damage and loss, or as stoppage of ventilation, especially in "Fire Stopes," would endanger many lives, the service is the most rigorous imaginable, demanding the highest type of reliability.

ARIZONA

FIRST Diesel installed 75 B.H.P., second 225 B.H.P. They operate at altitude of 5300 feet. Larger, operating triplex pumps, is direct connected to 150 K.W., 240 volt, D.C. generator. It runs under an average load of 400 amperes, 24 hours daily, making non-stop runs, amounting to forty days. This engine, in spite of the high altitude and an average load factor of 0.8, consumes only nine gallons fuel oil per hour, equivalent to 9.4 gallons per 100 K.W. hours, or 6.2 gallons per 100 B.H.P. hours assuming a generator efficiency of 87 per cent.

CONNECTICUT

TOTAL assets of this municipal plant $200,331.59, all paid for out of earnings, except $22,500, representing initial bonded debt. Never required help by taxation, although operating at lowest rates in the State. First Diesel bought in 1905, second 1907, third 1910, no steam equipment bought since first Diesel was installed. The General Superintendent, writing to prospective power purchasers, claims for his Diesels: quick response to variations in load, taking full rating without apparent effort: regulation as close as in the best steam practice; thorough reliability; dependable for long service runs; and, that his Diesels, carrying the bulk of the load, operate with gratifying fuel efficiency.

FLORIDA

EIGHT, twin triple-cylinder, engines, 450 B.H.P. each. They operate the largest phosphate mines in Florida. They say the Diesel makes oil the cheapest fuel in this State.

SIX units, 225 B.H.P. each. Owners write: "No trouble maintaining uniform loads, or paralleling with other engines. Oil used less than guarantee."

"WE installed one 225 B.H.P. Diesel about a year ago, operate twenty-four hour service, shutting down Sundays for five hours, and in the year's run, our service has been shut down but once, for two hours. This speaks for itself. We have been pleasantly surprised at the performance of the engine, and the fuel consumption is below the guarantee of the builders." Since old steam plant was discarded by the receivers and the Diesel adopted, the company has made good. Cheap reliable Diesel power for the city developed good off-peak loads, requiring additional Diesel equipment. The power house was built of bricks from stack of old steam plant.

INDIANA

A POWER station supplying a city of 5,000 and four small towns within fourteen miles. Current used for lighting, power and in water works. The two engines shown are 225 B.H.P. each, direct connected to alternating current generators. A twenty-four hour service is maintained, one engine running continuously, the other on peak loads. Daily fuel consumption 200 gallons.

KANSAS

TEN thousand five hundred dollars represents the yearly saving in fuel of these Diesel Units over the old superseded steam equipment. A greater saving will be made when the third unit, now on order, is installed. The equipment consists of two 240 B.H.P. units, direct connected to generators, which furnish power to a 1,200 barrel mill and a 350,000 bushel elevator.

KANSAS

THIS Diesel consumes daily 45 gallons of fuel oil, whereas the old steam equipment consumed 280 gallons on the same schedule, under the same load conditions. Sixty-cycle alternating current is generated and used for lighting, miscellaneous power requirements, and for driving water works pumps. Its success has lead two other Kansas municipalities to install Diesel equipment.

LOUISIANA

THIS installation consists of two Diesel engines of 120 B.H.P. each, driving 60 cycle alternating current generators; 100 gallons of fuel oil consumed daily furnishing the city, of 5,000 inhabitants, with all its water, electric light and commercial service. The electrically driven pumps are shown below. Two neighboring cities after watching these engines operate for a year, installed Diesels in their plants—proof of Diesel economy and reliability.

MARYLAND

PIANOS, of one of the most noted makes, are made in this factory, operated by the Diesel unit illustrated. It has been operating steadily for the past eight years, developing 225 B.H.P. at a speed of 165 R.P.M. It is direct connected to a 150 K.W. direct current generator.

MASSACHUSETTS

THIS large cotton mill employs its 450 B.H.P. double unit on a mill drive. Direct connected to alternator, it has been in daily operation for three years and has proven a most reliable source of power, ready at a moment's notice. Its ability to carry a steady, heavy mill drag and its freedom from smoke and soot, makes the Diesel a very desirable prime mover for the operation of textile mills. It occupies small space, requires very little attendance, consumes its fuel in proportion to requirements, and has no external flame or fire about it.

MASSACHUSETTS

WE have been in closer touch with this Massachusetts plant, consisting of 900 B.H.P. in three Diesel units than with any other, each week receiving a full operating report. Average fuel consumption 8.24 gallons per 100 net K.W. hours—equal to 6 gallons per 100 net B.H.P. hours. We do all we can to co-operate with Diesel owners and operators by maintaining a staff of inspecting engineers.

MASSACHUSETTS

THIS municipal plant, in a city of 5,000 inhabitants, started in 1903 with two Diesels of 120 B.H.P. each, shown in foreground, adding one of 225 B.H.P. in 1906 and another of like size in 1911—which speaks well for the satisfaction this city finds in Diesel operation. They operate 60 cycle generators in parallel, having a combined capacity of 500 kilowatts. The plant has 460 customers, to whom it supplies light and power for 310 H.P. of motors. It serves the city free, and is a paying institution without any account being taken of its municipal load.

MASSACHUSETTS

THIS large Central Station furnishes steam heat and electricity for a city of 32,000 inhabitants. Since 1906 it has used a 450 B.H.P. unit, which drives a 60 cycle alternator in parallel with steam equipment during extreme peaks and in emergency service. Being able to assume full load in less than one minute, this unit gives freedom from surprises and a sense of security not enjoyed with other prime movers. No stand-by losses, no extra attendance. This is just one of numerous plants with heating problems which have installed Diesels as auxiliary to steam equipment.

MINNESOTA

THERE are three installations of Diesels in this State which are auxiliary to water power. Two installed on showing of first. Writing to parties investigating our claims an engineer of one of these plants wrote: "We think the Diesel is more durable and dependable than steam. In answer to second part, I beg to advise steam cost for fuel $26.00 for 24 hours; Diesel, $3.40 for fuel and $0.50 for lubricating oil on $\frac{3}{4}$ load. We do not now charge any more when using engine than for water power, 10 cents per K.W.; when using steam we always had to raise price to 12 cents and it was hard to get even at that rate."

MISSOURI

FIRST Diesel built in America and the first to be placed under regular commercial load, here or abroad, was installed in the Anheuser-Busch Brewery, Saint Louis, in 1898. It operated in the bottling department until 1911, when it was superseded by the larger modern units shown.

The brewery is using Diesels with conspicuous success in its ice and refrigeration plants throughout the country. We issue a special bulletin on the economy of this plant which we will send upon request.

MISSOURI

THIS little Diesel Central Station of two units of 120 B.H.P. each, located under the shadow of one of the big distributing stations of the great Keokuk Dam water power plant, is generating and selling electric current for less than its big rival and making money.

NEW HAMPSHIRE

WHEN THE RIVERS RUN DRY

THE manager of this plant in writing to the company which had sold the electrical apparatus used in conjunction with its Diesels stated: "I will say that as an auxiliary I consider the Diesel the best proposition that can be installed. *** We are operating our Diesel units in parallel with our Hydro-Electric plant, located seventeen miles away, and we have never had the slightest difficulty. Another nice feature, where there is sufficient storage capacity, is the fact that from a 300 K.W. Diesel set, you can generate at least 7,000 K.W. hours in a twenty-four hour run, holding back sufficient water to take care of the peak loads. *** We are buying two more engines, after having had nearly three years' experience with our other two."

NEW JERSEY

WOOLEN fabrics are manufactured in this mill, which is operated by the 170 B.H.P. Diesel engine shown. Transmission is by rope drive. The engine has been in service ten to twelve hours every week day for eight years, having been installed in 1905. Oil consumption for this period is calculated by the mill at less than 7 gallons per 100 net B.H.P. hours.

NEW MEXICO

IN the irrigated sections of the Southwest central station plants find a good business furnishing electric current to farmers to run their irrigation pumps. This is what the 450 B.H.P. plant shown does besides lighting the streets and furnishing current for commercial purposes in a rapidly growing city in New Mexico.

NEW YORK

AN ideal installation. It consists of two 120 B.H.P. Diesels installed in 1907 and a third of 225 B.H.P. in 1912. These engines operate 60 cycle alternators in parallel. Their exhaust is passed into pits for muffling and reaches the atmosphere through the vent shown. Diesel owners dare to operate without spare units, making long continuous runs. Many Diesel plants are operated without shutting down for periods of over six weeks. Massive construction, finest materials, best of workmanship—these are the Diesel's guarantee of reliability.

NEW YORK

WRITING a prospective purchaser of Diesels, the manager of this Company stated: "We have not been without use of the engine any hour when required. It has never given us any trouble, and requires scarcely any attention. The entire expense of maintenance would not be more than $200.00 for the seven years that we have had it, and of that $200.00 we have spent, we have got half the parts in stock now for emergency. When we need additional power there will be nothing considered except another Diesel."

NEW YORK

AN ice and refrigeration plant of 455 tons capacity equipped with six Diesel units aggregating 1245 B.H.P., two of which are belt connected to ammonia compressors driven at constant speed, while two compressors are driven by variable speed motors. Raw water block ice is manufactured. Plant operates at very low cost and with great economy in space. Send for bulletin on this plant.

NEW YORK

A DIESEL located in a piano factory and surrounded by inflammable material, in the heart of New York's congested fire district. 170 B.H.P. free from fire risk.

OHIO

THE installation of this 225 B.H.P. engine required no changes to the old steam plant with which it operates. As is often the case, the existing engine room proved large enough to accommodate it. In many plants, as in the largest in the country producing encaustic tiling, Diesels supplement steam, utilizing to the best and most economical advantage available or idle space. Owing to the high economy of the Diesel, it is generally found most economical to load it to full capacity and let it carry the brunt of the power burden, leaving the steam installation to make up balance of requirements or carry peak loads. This makes an excellent arrangement, sufficient power being developed by the steam plant to supply the requisite exhaust steam for winter heating or manufacturing requirements.

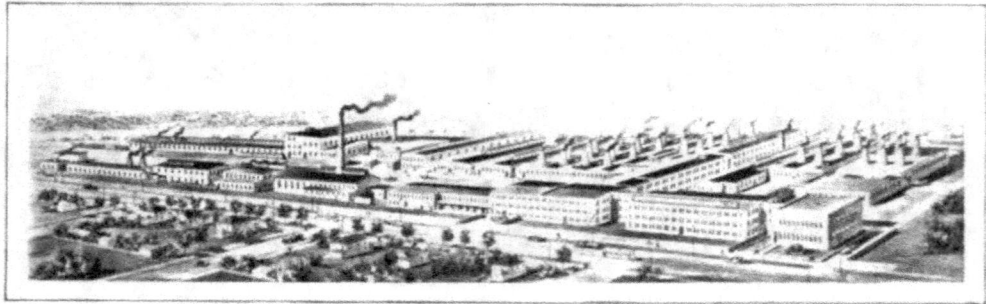

OHIO

FROM December 1905 to May first, 1906, according to published reports to tax payers, old steam plant caused a deficit of $6,714.49, met by taxation. From May twelfth, 1906, to October first, 1907, their two Diesels earned a surplus over all operating expenses of $3,928.71, thus turning a village liability into an asset. It was calculated by their engineer that during this period the Diesels saved $9,984.49 for fuel and $1,700.00 for day and night labor in boiler room—a total of $11,684.49.

OKLAHOMA

ONE of the most difficult and most dusty drives imaginable is that found in a cement manufacturing plant. Large heavy machinery such as rock crushers are constantly in use, causing heavy vibrations in the transmission and a very fluctuating load. In such plants extra precaution is taken in filtering the air which is admitted to the cylinders so as to prevent scoring and wear. This is the only precaution necessary on such a drive, the engine caring for this class of service perfectly. This installation, installed in 1912, consists of one 170 B.H.P. Diesel unit, belted to line shaft.

PENNSYLVANIA

THE three Diesels included in this plant, which manufactures city gas, are direct connected to alternating current generators. The power is used in manufacture and distribution by means of individual motor drives. This was one of the first plants in the country to use Diesel engines, having purchased two units of 75 H.P. in 1904, which are now supplemented by the modern units shown; in operation since 1907. ¶The Diesel was introduced into this plant, where gas was available at cost, solely because of its magnificent fuel economy.

PENNSYLVANIA

A MODERN Diesel installation in which the first two Diesels were installed in 1911 and additional units in 1912 and 1913. The plant now consists of six Diesels developing 1,350 B.H.P., operating crushers, grinders, chasers and finishers used in the manufacture of chocolate. Independent motor driven compressors have the advantage of great flexibility, being independent of the particular Diesel units in operation.

PENNSYLVANIA

THIS school for girls, which is located in the open country, belongs to the correctional system of institutions maintained by Pennsylvania. The state installed this Diesel of 120 B.H.P. on its showing as an economical power unit, which requires a minimum of attendance. This installation operates without regular attendance, generating a dependable direct current lighting and power supply to the entire satisfaction of the officials.

RHODE ISLAND

A 450 B.H.P. Diesel unit, direct connected to A. C. generator. In service eight years. No depreciation while idle, no stand-by expense. Starts any time in two or three minutes and maintains constant speed under variable loads. This unit used in the manufacture of silverware has been in constant service since 1905. It is operated in parallel with older steam equipment.

RHODE ISLAND

A CENTRAL Station in which Diesels began immediately to demonstrate their great economy. First Diesel installed in 1906. Use of steam discontinued and second Diesel installed 1907. Good profits and reliable service developing new business, three more were added in 1911. Plant now serves several neighboring towns and villages with 1125 B.H.P.

SOUTH DAKOTA

THIS plant operated one Diesel unit of 225 B.H.P. for one year. Then having become convinced of the Diesels advantages, reliability and economy, the owners added an additional unit of 225 B.H.P. It is one of numerous steam plants in which Diesels have superseded steam altogether.

TENNESSEE

THE municipal electric plant of a progressive little town of 3,000 inhabitants, which in 1912 installed its first Diesel of 225 B.H.P. direct connected to a 200 K.V.A. alternator. As can be seen, the city fathers wisely put their money into the engine instead of into the building.

TEXAS

A 675 B.H.P. central station. 3 phase, 2300-volt, 60-cycle. Average yearly cost of current 6.78 mills per K.W.H.—oil at $1.05 per bbl. Compressors direct coupled.

TEXAS

DIESELS are now operating several flour mills and elevators in different parts of the country with eminently satisfactory results. Economy of operation is the one big factor in such plants. In the modern mill shown, which began operation in 1912, the Diesel is connected to load by means of rope drives controlled by two friction clutches which render the different parts of the mill independent of one another. This feature is illustrated on page 34.

TEXAS

SPACE economy of Diesel installations strikingly shown. A belted 225 B.H.P. Diesel. Plant capacity: Refrigeration 35 tons, Ice 25 tons; Electric power 50 K.V.A.

TEXAS

THREE locomotive and car shops in this State have Diesel installations, operating traveling cranes, heavy machine tools and lighting buildings and yards. The plant illustrated installed two 120 B.H.P. Diesels in 1907, an additional unit of 225 B.H.P. being added in 1909.

The Superintendent of Motive Power and Rolling Stock of one of these plants wrote a brother official in another state: "I consider the Diesel Engine one of the most satisfactory and economical power plants obtainable. It is possible that the very close attention which this engine has received since it has been installed has resulted in its very successful operation both in point of economy as to fuel and repairs. The repairs which have been necessary are of so slight a nature that we have kept no special record as to the cost, but it is very low."

WISCONSIN

THE first municipal water and light plant to install Diesels in the world; operates four engines; 600 B.H.P. installed in 1905-11-13. Eighty arc lamps at $28.00 per lamp year cover operating cost—all other municipal lighting and 146,000,000 gallons pumpage free.

UNITED STATES NAVY

THE pontoon crane shown is used at a United States Navy Yard on the Atlantic Coast. The crane is of 150 tons capacity, and is operated by the 225 B.H.P. Diesel engine shown. This installation has been in successful operation one year. A duplicate of the above installation is serving the government at Pearl Harbor, Hawaii.

THIS COMPANY WISHES TO STATE THAT THE LARGE NUMBER OF ENGINES SOLD MAKES IT POSSIBLE TO RENDER TO ITS CUSTOMERS EXCEPTIONAL SERVICE. IT MAINTAINS A STAFF OF ERECTING AND OPERATING ENGINEERS OF LONG AND VARIED EXPERIENCE ENGAGED SOLELY IN VISITING OUR CUSTOMERS' INSTALLATIONS.

THE COMPANY MAINTAINS A CONSTANT INTEREST IN ALL DIESELS SOLD AND ENDEAVORS AT ALL TIMES TO SEE THAT THEY HAVE PROPER ATTENTION.

EVIDENCE

THE FOLLOWING STATEMENTS RELATIVE TO THE MERITS OF THE DIESEL ENGINE WERE MADE BY DIESEL OWNERS AND OPERATORS IN LETTERS WHICH THEY WROTE AT THE SOLICITATION OF PROSPECTIVE PURCHASERS OF POWER WHO WERE CONDUCTING THEIR OWN INVESTIGATIONS. THE STATEMENTS ARE CLASSIFIED UNDER FIFTEEN HEADINGS, VIZ.: ECONOMY, RELIABILITY, MAINTENANCE, REGULATION, OPERATION, LONG NON-STOP OPERATION, PARALLEL OPERATION AND INTERCHANGE OF CURRENT, OVERLOAD CAPACITY, CONSTRUCTION, REPAIRS, ATTENDANCE, MUNICIPAL INSTALLATION, DIESEL VERSUS STEAM AND WATER POWER, OUR CUSTOMERS RE-ORDER, SATISFACTION.

ECONOMY

"Our lubrication for two engines for 1911 was $472.75, and our cost per K.W., including repairs, fuel and lubrication, was $0.00253." *They operate two 225 B.H.P engines.*

❖ ❖ ❖

"As to comparative cost of operation, we figure that all things considered, we are operating for about 25 per cent. of what it would cost us to do the same work with steam. We figured that we could afford to junk our engines once every six years and come out better than even."

❖ ❖ ❖

"We have had one in use now upwards of seven years. It has given perfect satisfaction in every respect. The cost of repairs have been nominal, much less we believe than with a steam engine for the same length of time, and it has never been out of order except in one instance, where we were without its use for a few hours. The economy of operation of this engine is remarkable. We run our plant nine hours per day and we use anywhere from 63 to 70 gallons of oil per day, and our average load is fully 100 H.P."

❖ ❖ ❖

"We also find the fuel consumption to be almost in direct proportion to the work done; this, and the fact that there are no stand-by losses with this engine, have proved to be very important in our plant, as our load is a varying one and the peak load of short duration."

❖ ❖ ❖

"Our average running record for these engines, under all load conditions and averaged up by the month, is from $9\frac{1}{2}$ K.W. to 10 K.W. of electricity at the switchboard for every gallon of fuel oil consumed."

❖ ❖ ❖

"Our Chief Engineer states that the cost per H.P. per hour to operate the engines is about 2.07 mills. This is based on the price of Gas Oil at .0245 per gallon."

❖ ❖ ❖

"We clean the interior of the engines about twice a year and the work is done in the intervals when the engine would be shut down anyway. They require a little more careful attendance than a steam engine and perhaps a little more work to keep up, but we do not employ any more than we would have to operate the same number of steam units. The service is as reliable and satisfactory and our fuel cost is about 80 per cent. less than to develop the same horse-power with steam. For our use the Diesel has proven a complete success."

❖ ❖ ❖

"Our experience with these engines, under such long service, enables us to know them very thoroughly. Their regulation is fine; they are safe and reliable; the uniform efficiency and unvarying economy in the use of fuel oil gives them an excellent endorsement in this important factor."

❖ ❖ ❖

"The Diesels have given us very reliable and satisfactory service at an efficiency that is remarkable. We are highly pleased with them. The fuel consumption of the engines has never at any time exceeded the guarantee of the Company, and for our conditions we have saved about 85 per cent. of the cost of fuel over what our fuel would have been, using steam."

❖ ❖ ❖

"We certainly appreciate the reliability and economy of these engines. The fuel consumption being almost in direct proportion to the power delivered, thus enabling us to pull through our light load season with a profit, which we are certain could never have been accomplished with a steam plant."

❖ ❖ ❖

"The economy of these engines is no doubt ahead of anything on the market today."

❖ ❖ ❖

"Both engines are directly connected to alternating current generators from which we operate all pumps, air compressor, ice hoist and ice machine with motors. This gives a very flexible outfit, as we can operate installations in the plant with either or both engines as the case may require. We also furnish all city and commercial lights for the town and find that the engines come well within the guarantee of builders. We find that the engines are almost as efficient at half load as at full load so the fuel bill is in proportion to the power delivered."

❖ ❖ ❖

"We have found the oil consumption to be from 6 to $6\frac{1}{2}$ gallons per 100 B.H.P. hours."

RELIABILITY

"As you state, the fuel economy is conceded and we find in our experience that their reliability compares very favorably with other types of power. While we would not like to guarantee that any engine would give absolutely continual service, we feel that our experience is in favor of the oil engine in this respect, and if we were in your position would install them in preference to other power."

❖ ❖ ❖

"In our plant we have two of these engines, one of 225 H.P. and one of 170 H.P. We have never had any shut-down due to any fault of the engine, and as to our economy, they certainly have been the means of putting this plant on its feet."

❖ ❖ ❖

"We wired you on the 21st that we were about to contract for either a Diesel or————Oil Engine and asked you which you found to be the most reliable and gave the least trouble. We are in receipt of your wire stating that the Diesel was the most reliable and gave the least trouble, and was the least expensive to maintain. Please accept our thanks for your prompt reply. We closed a contract with the Diesel Engine Company for two of their 225 H.P. Engines to operate our mill. We were more favorably impressed with their engine from the start."

❖ ❖ ❖

"This plant has to operate all the time without shut-downs, except by accidents, which are extremely rare, as it not only supplies the entire commercial public lighting service of this city, but supplies electric power for the local manufacturing and industrial undertakings of a large scale—a large majority of which have no other source of power."

❖ ❖ ❖

"We have experienced no inconvenience or delay from any failure on the part of these engines. I do not hesitate in recommending them to anyone for any class of service whatsoever, provided, however, that they do not expect the engine to pull more than its rated H.P. We have, however, at times run our engines over-loaded, but I do not consider such operation good practice. The only precaution I would suggest, is to have a first-class engineer in charge of the Diesel Engine, and this however, I think is applicable to all engines."

MAINTENANCE

"As regards breakage, it has been very slight in our plant, having never experienced a shut-down on account of any fault of the engine. It is our practice to use only one man on a shift, although the Chief Engineer who has other duties is about the plant more or less during the day time. And when taking up bearings, etc, a helper from another part of the plant is called to assist."

❖   ❖   ❖

"During the year 1911 our average cost per K.W.H., including fuel, lubrication and repairs, was less than $0.003 (three mills) per K.W."

❖   ❖   ❖

"We have experienced very little difficulty in keeping these engines in good operative condition and the writer being a practical engineer had rather take care of the Diesel than steam engines. Our experience has been that the upkeep of these engines is nothing out of the ordinary."

❖   ❖   ❖

"Our maintenance cost has been $30.00 to $35.00 per engine per year and we have never had to renew any large parts on the engine, only piston rings, needle valves, springs and such small parts. Neither engine has even broken a valve spring for 18 months."

❖   ❖   ❖

"The upkeep of these engines we find to be only a very little more than that of a first-class steam equipment, while the economy is far superior, costing less than half that of steam."

❖   ❖   ❖

"In our opinion these engines are as reliable as the steam engine and we have experienced no unnecessary delays in service due to their failure. Our properties have been operating a Diesel plant at Sherman, Texas, for *about seven years*, and the first engines installed there are now doing as good service as new ones. The upkeep of the old engines last year in that plant amounted to less than 2 per cent. of the original cost."

❖   ❖   ❖

"In a report from the office of the City Clerk of a city in the State of Wisconsin, we find this item: *'Repairs on engines, average for 5 years, $125.00.'* This plant operates three Diesel Engines."

REGULATION

"The engines are used for driving three phase alternators for city lighting and power work and we find that they develop their rated H.P. to indicating instruments on the switchboards, and at no time have they exceeded the fuel consumption stated by the manufacturers. The speed regulation is good and the service reliable, in the years that we have had the engines. The engines need somewhat more careful attention than is usually given a steam engine, but there is no reason why an attendant who is even reasonably diligent can not operate the engines without trouble.

❖ ❖ ❖

"You ask—'Is the regulation as good as that of the ordinary Corliss Engine?' Yes, we consider it fully as good. The engines regulate entirely automatically, no throttling, but simply by a sensitive governor which varies the fuel supply to meet the exact load and speed requirements."

❖ ❖ ❖

"They are also very regular in speed, being perfectly adaptable to the operation of electric generators—for which purpose they are used in this plant, in connection with a number of steam engines."

❖ ❖ ❖

"When the last Diesel was installed, an extra foundation was built ready for another. The Diesel regulation also is excellent, and as for stopping from sudden breakage of its parts, it certainly will stop if things are not right, which is not to its discredit, but it is as reliable as good steam practice in every day service. Because the Diesel Engine embraces so many time saving advantages, is so safe to run and in many ways quite simple, it sometimes suffers from neglect, but if it is in good condition it will deliver its rated horse-power easily."

❖ ❖ ❖

"As to variations in motor load, they are considerable, especially when power and lighting lap, but we find the Diesels just as quick to respond as the steam engines. Roughly speaking, the load factor of the Diesels is about 75 per cent. with quite wide extremes during a 24-hour run according to demands, and they take up their full rating without apparent effort under proper operating conditions.

OPERATION

"Our Diesel engines carry the bulk of our load, though we have five steam engines which fill in on peak loads and extra service as required. All of our engines are directly connected to generators and the long runs are considerable, as the plant operates all the time, and, aside from a heavy commercial lighting load, has a connected load of about 1100 electric motors in the local factories, in addition to the street lighting service."

❖ ❖ ❖

"Our first engine was installed in February, 1906, and has been in operation ever since, running nights only. We have this year completed the installation of the second unit of 225 H.P. and believe that our new engine will fully come up to the excellent standard of service which the first one has given us."

❖ ❖ ❖

"Our Diesel engine handles this Ice Compressor, which is of 20 ton capacity, and in addition carries an electric load of about 300 amperes at 220 volts. We are satisfied that the engine is developing its full rated H.P. and the speed regulation for the 24 hours is so perfect as to procure the very best results from the Ice Machine."

LONG NON-STOP OPERATION

"Since that time, some eighteen months, we have depended absolutely on one engine to furnish light and power on a 24-hour schedule, allowing a weekly shut-down of about five hours on Sunday."

❖ ❖ ❖

"At the time we bought our first engine we were also interested in the ——— engine, but they would only give us a guarantee of 36 hours continuous running. It would take that long for the combustion chamber to fill with carbon and then the engine would have to be shut down and this chamber replaced with a clean one."

❖ ❖ ❖

"On one occasion we ran the engine six weeks, day and night, without stopping once."

❖ ❖ ❖

"From our experience with these engines, we can recommend them for the service of which you speak. We have in cases of necessity, run one of these engines six weeks continually without ever stopping it, although this is not a good policy with any engine, as they should be shut down and examined once a week at least."

PARALLEL OPERATION AND INTERCHANGE OF CURRENT

"Replying to your inquiry with reference to Diesel engines, have to say that on January 13, 1911, we started our Diesel Engines Nos. 293 and 294, which are three cylinder 16 inches by 24 inches direct connected to General Electric Company, three phase alternators. We have had no trouble even at the first trial to parallel these generators nor have we had any trouble at any time since then in putting them in parallel, and as to interchange of current between machines, it all depends upon the management of the engines. If furnished fair fuel and valves are kept in proper condition the interchange of current is negligible, and I consider the parallel operation of the Diesel thoroughly established and successful."

"We operate these units either singly or altogether, as the occasion demands, and we experience no difficulty in keeping our alternators in parallel."

"Replying to yours of the 12th inst., with reference to Diesel engines; beg to advise you that we are at present operating three of the 250 H.P. units directly connected to 187 K.V.A., 60 cycle, 2 phase alternators, running at 164 revolutions per minute. We are operating these units in parallel and are getting most satisfactory results from same."

OVERLOAD CAPACITY

"It might be well to state that while this engine is rated at 225 H.P., on Saturday night, July 6, 1912, the engineer informs me that it was up to 250 H.P."

"Our small engine has been operating at an overload much of the time for the past year and for this reason we have purchased additional (*Diesel*) equipment."

"The writer some years ago had charge of a plant which was equipped with Diesel engines, direct connected to generator, and it was our practice there to carry 150 K.W., on the switchboard, with the 225 H.P. engine when it was carrying its own compressor, although at times we were obliged to carry as high as 190 K.W."

CONSTRUCTION

"When building the foundation for the engine the management did not think it necessary to follow the plans, but followed their own ideas. Instead of excavating to solid rock, they were satisfied to build the foundation on rotten surface rock. The foundation was made of concrete. Sometime after starting the engine, same was found to be in motion and the concrete foundation also, wobbling up and down like on a pivot. This increased with time to pretty near an inch and brought the engine out of line with air compressor. The 170 H.P. engine, weighing about 34,000 pounds, caused a pressure and vibration in the rotten rock underneath, whereby same was disrupted and crushed. This mistake we have now corrected at an expense of over $500.00, when it would not have cost more than $50.00 if the plans had been followed. It is a wonder to us here that the engine, with all its fine mechanism, did not fall to pieces. We have had enough of steam and would not trade our Diesel for all the steam engines there are."

❖ ❖ ❖

"The engines are well built and are very massive in construction, and after six years of continuous work our oldest engine is still giving perfect satisfaction. They wear well and we do not find that the cost of upkeep amounts to more than that of any first-class steam engine."

REPAIRS

"We have had one in use now seven years and it has been perfectly satisfactory in every respect. It requires scarcely no attention whatever, and our repair bills or expenditures for new parts have been very small indeed and we have not been without the use of the engine any hour when it was required."

❖ ❖ ❖

"It is rather difficult to give you an approximate yearly repair cost as this has varied with us according to the work done and parts replaced. Last year our repairs did not foot up to $50.00, but a year ago were over $300.00. Even if an unexpected accident should make the repair cost exceedingly high, the low cost of operation will make you the gainer in the long run."

❖ ❖ ❖

"We have not found the repair cost on them any higher than might be expected on the whole of a steam equipment of like power."

See also under MAINTENANCE

ATTENDANCE

"We do not find that it requires an expert to operate these machines but we do advise that care should be taken in selecting a man of ordinary intelligence, who is careful, trustworthy and faithful."

❖ ❖ ❖

"There is no reason why one man should not operate two engines with their generators and the switchboard, that is while running. Of course when there is adjustment, cleaning or repairs to be done more help will be necessary. There are no jobs about starting or running the engine that one man cannot do."

❖ ❖ ❖

"Will state that we think the best recommendation that we can give them is the fact that we are now installing our third unit after having one in use for nearly five years. In the engine room we have two assistant engineers and one chief. It is their duty to look out for the machinery of the electrical plant as well as a 20 ton ice plant, which we also operate. We are now increasing the ice plant to a 40 ton output and the same crew will be able to care for it. One man runs the whole outfit at night."

❖ ❖ ❖

"It has been my experience that, with the great economy of these engines, we can well afford to pay the price for a good man; we are then taking no chance of a cheap man destroying a high priced machine, and, in the end, we are way ahead of the game over a steam driven plant."

❖ ❖ ❖

"There is practically no labor required in the operation of these engines."

❖ ❖ ❖

"You should have an engineer of some intelligence to take care of the engines and keep them up, the same as with any first-class steam engine, but, after once started, the engines are almost automatic, requiring only an attendant to watch them and see that they get proper lubrication. We do not find that their upkeep is any greater than a steam plant. Of course, engines in duplicate guard against any shut-down in case of accident, but we ran one unit for more than two years and are satisfied that it gave as good, if not better, service than any single unit steam plant in the state operating continuously. On one occasion we ran the engine six weeks, day and night, without stopping once."

MUNICIPAL INSTALLATION

"*** This is an actual monthly report of the electric light and waterworks of this city, showing a cost of 8 mills per K. W. hour on the switchboard. The waterworks pumps are motor driven in duplicate. We sell commercial service enough to make the street lighting and water service free to the city." *And again:*

"*** This plant pays. Come and see us, we like to show what we have. The above is an actual report of the total cost of operating the water and lighting plant of this city with Diesel engines, which is approximately $28.00 per lamp year for 80 arc lamps, with 146,000,000 gallons pumpage gratis."

❖ ❖ ❖

"In our opinion Municipal Ownership of the Public Utilities, particularly the lighting system in this case, is the best possible solution of the problem. But a short time ago a proposition was made in the Village by a corporation, to purchase the Municipal Lighting Plant, but on being placed before the people was beat nearly two to one. A year later the proposition was brought up for additional power in the Municipal System and was carried by a vote of about 200 to 5. This is the best testimonial which we can give for the Municipal Ownership. You can easily see that this is what we would advise."

❖ ❖ ❖

"To argue these matters in a letter is almost impossible and I would therefore advise you to have your committee come here and look over our plant and see our Diesel engine. I think it would be more profitable for your city to install a Diesel engine and have a plant of your own."

❖ ❖ ❖

"Now as to our plant; it was built in 1892 and will complete a continuous record of success next month, covering a period of 20 years. It represents an investment of about $200,000, of which all but $22,500 has been paid from profits, since commercial lighting and power were added to the original street-lighting plant in 1898, and at the present rate the entire debt will be wiped out within a year. Our rates have always been the lowest in this state, and nearly all the factories of this city are operated by the power of this plant—about 1200 connected H.P. in motors—not to mention a heavy lighting load. As to fuller details, we take pleasure in sending you our last annual report of nearly a year ago under separate cover, and trust that we will thus supply you with the information desired." *This plant installed its first Diesel in 1905 and has added no additional steam equipment since then.*

DIESEL vs. STEAM AND WATER POWER

"Another great advantage of the Diesel engine over steam is that we do not have to wait half a day or more for results; we can start our engine in less than three minutes, and pull the load from the start. As to maintenance, we can find no reason why the expense should be very great. Our expenses with the old steam engine were numerous and heavy. The engine is durable—there is no question about that."

❖ ❖ ❖

"You say you are now operating steam engines and think of buying a new engine. We have gone through the same ordeal and hesitated and figured and figured and hesitated and studied up on different kinds of engines, so I know how you feel. We would not here go back to the old steam or have any other power next to our water-power than the Diesel engine."

❖ ❖ ❖

"Right here I would like to say that about the time of my taking charge of this plant, the financial condition of the Company was such that if we had a steam plant we could never have pulled through, and, when in need of more power, we will install another Diesel."

❖ ❖ ❖

"We have found the Diesel engine remarkably efficient in the use of fuel oil, having placed our first one in service in 1905, another in 1907 and the last one in 1910, with a foundation ready for still another, so that our experience has been considerable. We also have steam engines in service, but have not added to our steam equipment since the first Diesel was installed."

❖ ❖ ❖

"After running this engine for two years, our business had increased so that it was necessary to purchase a second engine of the same size, and this Spring we installed a third unit of 225 H.P. When installing the first engine we discarded a steam outfit, and the fact that we have continued to buy Diesel engines as our business increased, should convince you of our faith in them."

❖ ❖ ❖

"We will take our Diesel in preference to steam every time and if you get the engines manufactured by the Busch-Sulzer Bros.-Diesel Engine Co., St. Louis, we know that they will deliver the goods every time if you take care of them. You can also rest assured that these engines will fulfill every guarantee made by their builders."

OUR CUSTOMERS RE-ORDER

"We have been using Diesel engines in our plant for the past six years and we are well pleased with the service they have given. Our first engine was of 170 H.P. and was the only unit we had for three years. At the end of that time business had increased to such an extent that we found it necessary to duplicate our outfit with another 170 H.P. unit. We continued to grow, and last year installed the third unit of 225 H.P."

❖ ❖ ❖

A superintendent writes to his management: "The present engine is doing very good work, is not giving any trouble at all since we fixed up the starting cam, four months ago, and I still believe the Diesel unit is the most reliable, as well as the most economical outfit, that ever generated current. I am decidedly of the opinion that the installation of another engine is a step in the right direction and will effect large economies in our operation."

❖ ❖ ❖

"When our load became so great that we were unable to handle it with an engine of this size, we immediately installed a second machine of 225 H.P., which has been giving excellent service since March 1st last, when it was placed in service. This fact answers your question as to whether we would buy Diesel equipment if building again; we most certainly should Mr. ——, and have."

❖ ❖ ❖

"As to our experience with the Diesel engine, we have much to commend and very little to say of a negative nature regarding it; the fact that we have purchased three of them in succession and have a foundation built for another, speaks of what our faith has been in them."

❖ ❖ ❖

"We have now ordered a third and larger engine which will be installed next February, and while we are building we are making our power house large enough to accommodate the fourth unit when it is needed. We think they are the greatest thing out."

❖ ❖ ❖

From telegram: "Sixty K.W. generator over six years, at 10 hours full load and over, fuel consumption less than Diesel Company's guarantee. Besides two engines seventy-five each, seven years, we have one two hundred twenty-five, running five months very satisfactorily."

SATISFACTION

Every extract exhibited in EVIDENCE testifies to the satisfaction Diesel users are getting from their engines—but here are some more:

"We further have this to say—that the Diesel Oil Engine has made it possible for this plant to succeed. Our experience with this engine compels us to speak very highly of it, and we do not hesitate to say that we believe it to be the most economical and reliable engine on the market today."

❖ ❖ ❖

"I might also add that Mr. ——————, President and Manager of the —————— Ice, Light & Power Co., told me in a conversation a few days ago, that the results obtained from the Diesel engine were away beyond his highest expectations, and that he intended installing them in several of the electrical plants which he owns."

❖ ❖ ❖

"We have had enough of steam and would not trade our Diesel for all the steam engines there are. You can tell the people of Belleville that we believe the Diesel engine to be the best and cheapest power produced in the world today, as I see from your letter that your intention is to install a municipal plant. Corporations have tried to buy our plant, but we are glad that we were not ensnared."

❖ ❖ ❖

"As to your natural inquiry—'Is the engine entirely satisfactory?'—we will simply say that our reason for adopting it was because we were looking for something more satisfactory than steam, although we already had an excellent steam plant. Now we are looking for something better than the Diesel and if there is anything that will beat it in fuel economy, speed regulation, safety and many other essential features, we would like to know where to find it."

❖ ❖ ❖

"These engines are both 3-cylinder engines of the 4-cycle type, one of 170 and one of 225 B.H.P., which have been in operation for the last seven years, and are still giving perfect satisfaction. We are highly pleased with these engines, both as to reliability and economy; in fact in every respect. We can certainly recommend this engine very highly and suggest, before purchasing your prime mover you write to the makers of this engine for detailed information."

ENGINEERING DATA

ON

POWER FACTOR
ATMOSPHERE
EFFECT OF ALTITUDE
MINERAL OILS
BEAUMÉ SCALE
CONCRETE FOUNDATIONS
LEATHER BELTING
UNIT EQUIVALENT IN OTHER UNITS
CAPACITY OF CYLINDRICAL TANKS
TABLE FOR EQUALIZING PIPES
WATER
WEIGHT OF MATERIALS

POWER FACTOR

On an alternating-current electric circuit, the product of the readings obtained simultaneously from a volt-meter and an ammeter indicating the apparent power, may be more than the reading obtained at the same time on a wattmeter which indicates the true power. The power factor is the ratio of the wattmeter reading to the product of the voltmeter and ammeter readings and is never greater than one. In any case the power factor is the ratio of true power to apparent power. This ratio is usually expressed in percent and can never be greater than 100 per cent. If true power is expressed in kilowatts (kw.) and apparent power expressed as the product of kilovolt amperes (kva.), then the following formula can be used:

$$\text{Power Factor (P. F.) equals } \frac{kw.}{kva.}$$

For estimating purposes, the following may be assumed as average values of power factors in their respective circuits: Incandescent lighting load, no motors, 95 per cent.; Incandescent lighting and induction motors, 85 per cent.; induction motors only, 80 per cent.; arc lamps 70 per cent.

The true power, in kw., equals the average volts between line terminals, multiplied by the average amperes line current, multiplied by the power factor (expressed as a decimal fraction), divided by 1000, and multiplied by:

- 1 for single phase
- 2 for two phase
- 1.732 for three phase

Although the current, equivalent to the difference between the apparent and true powers, imposes practically no load upon the prime mover (engine); this, so called "wattless current," produces in the generator a heating greater than that due to the equivalent true power, and the generator must, therefore, be proportioned to take care of this current without over-heating.

❖   ❖   ❖

ATMOSPHERE
(Atmospheric pressures)

One atmosphere (based on sea-level) equals 14.7 pounds per square inch
"   "   "   "   "   " 29.922 inches of mercury.
"   "   "   "   "   " 33.9 feet of water.

EFFECT OF ALTITUDE

Table of Altitudes in feet above sea-level; with corresponding approximate Barometric Readings, Atmospheric Pressures and proportionate Densities.

(The capacity of an internal combustion engine at higher altitudes, as compared with its capacity at sea-level, is practically proportional to the atmospheric densities.)

Altitude in Feet	Barometer in Inches	Atmospheric Pressure in pounds per square inch	Proportionate Atmospheric Density
0.00	30.0	14.72	1.00
500.	29.5	14.45	0.98
1000.	28.9	14.18	0.96
1500.	28.4	13.94	0.94
2000.	27.9	13.69	0.93
2500.	27.4	13.45	0.91
3000.	26.9	13.20	0.89
4000.	26.0	12.75	0.86
5000.	25.1	12.30	0.83
6000.	24.2	11.85	0.80
7000.	23.3	11.44	0.77
8000.	22.5	11.04	0.75
9000.	21.7	10.65	0.73
10000.	20.9	10.26	0.70

❖ ❖ ❖

MINERAL OILS

The characteristics of Crude Mineral Oils and their products vary greatly in different localities; but the following general information may be of interest.

	Gravity, deg. Bé	Flash Point deg. F.	Burning Point deg. F.
Crude Oil	12 to 45	110 to 200	120 to 220
Kerosene	40 to 50	90 to 125	105 to 150
Distillate (Gas Oil)	28 to 38	100 to 250	110 to 325
Fuel Oil	22 to 28	100 to 300	125 to 375
Residuum	10 to 20	125 to 500	200 to 600

The heat value of mineral oils and their products may be very closely determined from their gravity, by the following formula:

$$\text{B.T.U. per pound} = 18650 - \{40 (\text{Beaumé} - 10)\}$$
(Sherman & Krapff)

"Asphaltum," as applied to a constituent of some mineral oils, is a most indefinite term, as its definition and the method of its determination have not been standardized. Until some standard is agreed upon, it would be better to compare "Asphaltum base oils," for use in Diesel engines, on the basis of the percentage of weight remaining after reduction to constant weight in a closed furnace at a definite temperature, say 300 deg. Centigrade.

BEAUMÉ SCALE

The density of crude or fuel oil is usually specified in "degrees Bé", at 60 degrees F.

The Beaumé hydrometer is an instrument for determining the density of liquids. The graduations are in numbers, termed "degrees", of an arbitrary scale.

LIQUIDS LIGHTER THAN WATER

Degrees Beaumé	Specific Gravity	Weight of One Gallon, Pounds	Degrees Beaumé	Specific Gravity	Weight of One Gallon, Pounds
10.0	1.	8.33	27.0	0.892	7.44
11.0	0.993	8.28	28.0	0.886	7.39
12.0	0.986	8.22	29.0	0.881	7.34
13.0	0.979	8.16	30.0	0.875	7.30
14.0	0.972	8.10	31.0	0.870	7.25
15.0	0.966	8.05	32.0	0.864	7.21
16.0	0.959	7.99	33.0	0.859	7.17
17.0	0.952	7.94	34.0	0.854	7.11
18.0	0.946	7.88	35.0	0.849	7.07
19.0	0.940	7.83	36.0	0.843	7.02
20.0	0.933	7.78	38.0	0.833	6.94
21.0	0.927	7.73	40.0	0.824	6.86
22.0	0.921	7.68	42.0	0.814	6.78
23.0	0.915	7.63	44.0	0.805	6.71
24.0	0.909	7.58	46.0	0.796	6.63
25.0	0.903	7.53	48.0	0.787	6.56
26.0	0.897	7.48	50.0	0.778	6.48

❖ ❖ ❖

CONCRETE FOUNDATIONS

Concrete for foundations to be 1 part Portland Cement, 2½ parts clean, sharp sand, and 5 parts clean, broken stone. Stone to pass through a 2-inch ring. To be mixed wet, and rammed every 8-inch depth until water appears on surface. Templets to be made open to permit ramming. Concrete must not be allowed to set hard before other concrete is placed on it, otherwise sound bonding between portions cannot be had. Around each fundation bolt a wooden box, 4 inches square at lower end, with increasing taper of ¼-inch per foot, and at least 4 feet long, must be placed. As the concrete sets boxes to be rapped loose and withdrawn. If preferred, 4-inch diameter galvanized spouting may be used and left in foundation, projecting not more than ½-inch above concrete. Grouting to be equal parts Portland Cement and clean sand, mixed wet enough to flow readily. Grouting must fill spaces around foundation bolts.

LEATHER BELTING

The size of leather belting, suitable for any given work, depends upon so many factors that it is practically impossible to prepare a simple table which will meet all requirements. The table given below, however, is safe for all ordinary conditions.

HORSE-POWER PER ONE-INCH WIDTH OF BELT

Belt Speed Feet per Second	Endless, Cemented			Lapped and Riveted			Lapped and Laced		
	Single	Double	Triple	Single	Double	Triple	Single	Double	Triple
10	0.73	1.33	1.94	0.61	1.12	1.63	0.52	0.96	1.39
20	1.45	2.65	3.85	1.21	2.22	3.23	1.02	1.88	2.74
30	2.15	3.92	5.70	1.79	3.28	4.78	1.52	2.78	4.05
40	2.80	5.12	7.45	2.34	4.30	6.25	1.97	3.60	5.25
50	3.42	6.25	9.10	2.83	5.20	7.55	2.38	4.37	6.35
60	3.98	7.25	10.60	3.28	6.00	8.75	2.74	5.00	7.30
70	4.45	8.15	11.85	3.66	6.35	9.75	3.02	5.55	8.05
80	4.70	9.00	13.05	3.95	7.25	10.60	3.22	5.80	8.60
90	5.20	9.60	13.90	4.20	7.70	11.15	3.35	6.15	8.95
100	5.45	10.00	14.50	4.30	7.90	11.45	3.37	6.20	9.00

The thickness of belt assumed is:

    Single Belt— $3/16$-in. minimum.
    Double belt— $11/32$-in. minimum.
    Triple Belt— $1/2$-in. minimum.

For special thickness the tabled figures may be proportionately corrected.

The most satisfactory belt speed, all things considered, is between 60 and 70 feet per second, although the most economical speed is about 100 feet per second, which is, however, too high for ordinary iron pulleys.

The belt should, if possible, be arranged to have the tight, or pulling, side on the bottom.

The pulley ratio (the proportion of driving to driven pulley diameters, or vice versa) should not be greater than 6 to 1.

The distance between pulley centers should vary with the thickness and width of the belt, and the pulley ratio. No definite rules for this have ever been formulated, and there is considerable diversity of opinion. It will, however, be found advisable to make the distance between pulley centers NOT

LESS THAN proportionate to the following table, for a pulley ratio of 1 to 1. For a pulley ratio of 6 to 1 this distance should be increased 20 per cent., and proportionately between the 1 to 1 and the 6 to 1 ratios.

SINGLE BELT	DOUBLE BELT	TRIPLE BELT
3 in. wide– 5 ft. centers	6 in. wide– 8½ ft. centers	12 in. wide–13 ft. centers
6 " " 7½ " "	12 " " 11½ " "	24 " " 18 " "
12 " " 10 " "	24 " " 16 " "	48 " " 25 " "

In ordering a belt it is well to inform the belt manufacturer of the following conditions, and to require him to furnish a guarantee that the belt will satisfactorily perform the required work under the stated conditions:

> Horse-power to be transmitted.
> Speed and size of driving pulley.
> Speed and size of driven pulley.
> Distance between pulley centers.
> Height above floor line of driving pulley.
> Height above floor line of driven pulley.
> Direction of rotation of driving pulley.
> Direction of rotation of driven pulley.
> Whether locality is dry or damp.

Usually a sketch will give the above information more clearly than a written description.

❖   ❖   ❖

UNIT EQUIVALENT IN OTHER UNITS

1 Horse-Power (H.P.) equals
- 746 watts.
- 0.746 K.W.
- 33,000 ft.-lbs. per minute.
- 550 ft.-lbs. per second.
- 2,545 heat-units per hour.
- 42.4 heat-units per minute.
- 0.707 heat-units per second.

1 Kilowatt (K.W.) equals
- 1,000 watts.
- 1.34 H.P.
- 2,654,200 ft.-lbs. per hour.
- 44,240 ft.-lbs. per minute.
- 737.3 ft.-lbs. per second.
- 3,412 heat-units per hour.
- 56.9 heat-units per minute.
- 0.948 heat-units per second.

1 Heat-unit (B.T.U.) equals
- 1,055 watt seconds.
- 778 ft.-lbs.
- 0.000293 K.W. hour.
- 0.000393 H.P. hour.

CAPACITY OF CYLINDRICAL TANKS

Diameter in Feet and Inches, Area in Square Feet and U. S. Gallons per Foot in Length or Depth.

1 U. S. gallon = 231 Cu. in. = 0.13368 Cu. ft.

Diameter Ft. In.	Area Sq. Ft.	Gallons per 1 ft.	Diameter Ft. In.	Area Sq. Ft.	Gallons per 1 ft.
2	3.142	23.50	7 - 6	44.18	330.48
2 - 6	4.909	36.72	8	50.27	376.01
3	7.069	52.88	8 - 6	56.75	424.48
3 - 6	9.621	71.97	9	63.62	475.89
4	12.566	94.00	9 - 6	70.88	530.24
4 - 6	15.90	118.97	10	78.54	587.52
5	19.63	146.88	12	113.10	846.03
5 - 6	23.76	177.72	15	176.71	1321.90
6	28.27	211.51	20	314.16	2350.10
6 - 6	33.18	248.23	25	490.87	3672.00
7	38.48	287.88			

❖ ❖ ❖

TABLE FOR EQUALIZING PIPES

The size of main pipe is given in the column at the left. The number of branches is given in the line on top, and the proper size of branches is given in the body of the table on the line of each main and beneath the desired number of branches.

In commercial sizes the nominal 1¼-inch pipe is generally over-size; often as large as 1⅜. It is safe to call it 1.3 inches, and it is so figured in the table. Exact sizes are given for branch pipes. The designer of the pipe system can thus better select the commercial sizes to be used.

Size of Main Pipe	\multicolumn{15}{c}{NUMBER OF BRANCHES.}
	2	3	4	5	6	7	8	9	10	11	12	13	14	15	16
1 In.	.758	.644	.574	.525	.488	.459	.435	.415	.398	.383	.370	.358	.348	.338	.330
1¼ "	.985	.838	.747	.683	.635	.597	.556	.540	.518	.498	.482	.466	.452	.440	.428
1½ "	1.14	.967	.861	.788	.733	.689	.653	.623	.597	.575	.555	.538	.522	.508	.494
2 "	1.52	1.29	1.15	1.05	.977	.918	.870	.830	.796	.766	.740	.717	.696	.677	.660
2½ "	1.89	1.61	1.44	1.31	1.22	1.15	1.09	1.03	.995	.958	.925	.896	.870	.846	.825
3 "	2.27	1.92	1.72	1.58	1.47	1.38	1.31	1.25	1.19	1.15	1.11	1.08	1.04	1.02	.989
3½ "	2.65	2.26	2.01	1.84	1.71	1.61	1.52	1.45	1.39	1.34	1.30	1.25	1.22	1.18	1.15
4 "	3.03	2.58	2.30	2.10	1.95	1.84	1.74	1.66	1.59	1.53	1.48	1.43	1.39	1.35	1.32
4½ "	3.41	2.90	2.58	2.36	2.20	2.07	1.96	1.87	1.79	1.72	1.67	1.61	1.57	1.52	1.48
5 "	3.79	3.22	2.87	2.63	2.44	2.30	2.18	2.08	1.99	1.92	1.85	1.79	1.74	1.69	1.65
6 "	4.55	3.87	3.45	3.15	2.93	2.75	2.61	2.49	2.39	2.30	2.22	2.15	2.09	2.03	1.98
7 "	5.30	4.51	4.02	3.68	3.42	3.21	3.05	2.91	2.79	2.68	2.59	2.51	2.44	2.37	2.31
8 "	6.06	5.16	4.59	4.20	3.91	3.67	3.48	3.32	3.18	3.09	2.96	2.87	2.78	2.71	2.64
9 "	6.82	5.80	5.17	4.73	4.40	4.13	3.92	3.74	3.58	3.45	3.33	3.23	3.13	3.04	2.97
10 "	7.58	6.44	5.74	5.25	4.88	4.59	4.35	4.15	3.98	3.83	3.70	3.59	3.48	3.38	3.30
12 "	9.08	7.73	6.89	6.30	5.86	5.51	5.22	4.98	4.78	4.60	4.44	4.30	4.18	4.06	3.96

WATER

One cubic foot of water, weighs 62.425 pounds.
" " " " " equals 7.48 U. S. gallons.
One pound of water, equals 27.7 cubic inches.
One U. S. gallon of water, weighs 8.331 pounds.
" " " " " equals 231 cubic inches.
" " " " " " 0.1134 cubic feet.
One foot head of water, equals 0.4335 pounds per square inch.
One pound per square inch equals 2.307 feet head of water.
One foot head of water equals 0.8826 inches of mercury.
One inch column of mercury equals 1.133 feet of head of water.

The foregoing figures are for water at temperature of maximum density (39.1° F.); but are sufficiently close for all ordinary temperatures.

Sea water weighs about 2¾ per cent. more than fresh water.

❖ ❖ ❖

WEIGHT OF MATERIALS

Brass, average	0.301 pounds per cubic inch.
Bronze "	0.320 " " " "
Copper	0.320 " " " "
Cast Iron	0.260 " " " "
Wrought Iron	0.278 " " " "
Lead	0.411 " " " "
Steel	0.283 " " " "
Tin	0.265 " " " "
Zinc	0.253 " " " "
Alcohol	6.67 pounds per U. S. gallon.
Ammonia	7.50 " " " " "
Linseed Oil	7.84 " " " " "
Mineral Oil	6.5 to 8.1 pounds per U. S. gallon.
Cedar	39 pounds per cubic foot.
Hemlock	24 " " " "
Hickory	48 " " " "
Lignum Vitæ	62 " " " "
Mahogany	51 " " " "
Maple	42 " " " "
White Oak	48 " " " "
Red Oak	46 " " " "
White Pine	28 " " " "
Yellow Pine	38 " " " "
Spruce	28 " " " "
Common Brick	112 " " " "
Fire Brick	145 " " " "
Portland Cement	115 " " " "
Clay	140 " " " "
Concrete	145 " " " "
Ice	56 " " " "
Sand (dry)	100 " " " "
" (wet)	125 " " " "
Stone	135 to 200 " " " "
Soft Coal	55 " " " "
Hard Coal	60 " " " "
Coke	35 " " " "